CONCEPTS OF BIOPHYSICS

CONCEPTS OF BIOPHYSICS

Dr. Ashok Kumar

RANDOM

RANDOM PUBLICATIONS

NEW DELHI - 110 002 (INDIA)

Concepts of Biophysics

ISBN 978-93-51117-46-9

Published in 2015 in India by
RANDOM PUBLICATIONS
4376-A/4B, Gali Murari Lal, Ansari Road
New Delhi-110 002
Phone: +9111-43580356, 23289044
E-mail: randomexports@gmail.com; sales@randompublications.com; info@randompublications.com

Reprinted 2026

Type Setting by: Friends Media, Delhi-110089

Preface

Biology studies life in its variety and complexity. It describes how organisms go about getting food, communicating, sensing the environment, and reproducing. On the other hand, physics looks for mathematical laws of nature and makes detailed predictions about the forces that drive idealized systems. Spanning the distance between the complexity of life and the simplicity of physical laws is the challenge of biophysics. Looking for the patterns in life and analysing them with math and physics is a powerful way to gain insights. Biophysics looks for principles that describe patterns. If the principles are powerful, they make detailed predictions that can be tested. Biophysics spans all scales of biological organization, from the molecular scale to whole organisms and ecosystems. Biophysical research shares significant overlap with biochemistry, nanotechnology, bioengineering, computational biology and (complex) systems biology. It has been suggested as a bridge between biology and physics.

Molecular biophysics typically addresses biological questions similar to those in biochemistry and molecular biology, but more quantitatively. Scientists in this field conduct research concerned with understanding the interactions between the various systems of a cell, including the interactions between DNA, RNA and protein biosynthesis, as well as how these interactions are regulated. A great variety of techniques are used to answer these questions. Fluorescent imaging techniques, as well as electron microscopy, x-ray crystallography, NMR spectroscopy, atomic force microscopy (AFM) and small-angle scattering (SAS) both with X-rays and neutrons (SAXS/SANS) are often used to visualize structures of biological significance. Conformational change in structure can be measured using techniques such as dual polarisation interferometry, circular dichroism, SAXS and SANS. Direct manipulation of molecules using optical tweezers or AFM, can also be used to monitor biological events where forces and distances are at the nanoscale. Molecular biophysicists often consider complex biological events as systems of interacting entities which can be understood e.g. through statistical mechanics, thermodynamics and chemical kinetics. By drawing knowledge and experimental techniques from a wide variety of disciplines, biophysicists are often able to directly observe, model or even manipulate the structures and interactions of individual molecules or complexes of molecules. In addition to traditional (i.e. molecular and cellular) biophysical topics like structural biology or enzyme kinetics, modern biophysics encompasses an extraordinarily broad range of research, from bioelectronics to quantum biology involving both experimental and theoretical tools. It is becoming increasingly common for biophysicists to apply the models and experimental techniques derived from physics, as well as mathematics and statistics, to larger systems such as tissues, organs, populations and ecosystems. Biophysical models are used extensively in the study of electrical conduction in single neurons, as well as neural circuit analysis in both tissue and whole brain. Many biophysical techniques are unique to this field. Research efforts in biophysics are often initiated by scientists who were traditional physicists, chemists, and biologists by training.

This book provides the background needed for fundamental training in biophysics and offers a great deal of advanced biophysical knowledge. All graduate science students will find this book an accessible and valuable aid in learning how these concepts can be used in biological research.

I thank all members of my team who have helped in the preparation of the book. My special thanks go to "Random Publications" who have published the book.

— Dr. Ashok Kumar

Contents

Chapter 1

Introduction

Biophysics is an interdisciplinary science using methods of, and theories from, physics to study biological systems. Biophysics spans all scales of biological organization, from the molecular scale to whole organisms and ecosystems. Biophysical research shares significant overlap with biochemistry, nanotechnology, bioengineering, computational biology and (complex) systems biology. It has been suggested as a bridge between biology and physics.

The term "biophysics" was originally introduced by Karl Pearson in 1892.

Molecular biophysics typically addresses biological questions similar to those in biochemistry and molecular biology, but more quantitatively. Scientists in this field conduct research concerned with understanding the interactions between the various systems of a cell, including the interactions between DNA, RNA and protein biosynthesis, as well as how these interactions are regulated. A great variety of techniques are used to answer these questions.

Fluorescent imaging techniques, as well as electron microscopy, x-ray crystallography, NMR spectroscopy, atomic force microscopy (AFM) and small-angle scattering (SAS) both with X-rays and neutrons (SAXS/SANS) are often used to visualize structures of biological significance. Conformational change in structure can be measured using techniques such as dual polarisation interferometry, circular dichroism, SAXS and SANS. Direct manipulation of molecules using optical tweezers or AFM, can also be used to monitor biological events where forces and distances are at the nanoscale. Molecular biophysicists often consider complex biological events as systems of interacting entities which can be understood e.g. through statistical mechanics, thermodynamics and chemical kinetics. By drawing knowledge and experimental techniques from a wide variety of disciplines, biophysicists are often able to directly observe, model or even manipulate the structures and interactions of individual molecules or complexes of molecules.

In addition to traditional (i.e. molecular and cellular) biophysical topics like structural biology or enzyme kinetics, modern biophysics encompasses an extraordinarily broad range of research, from bioelectronics to quantum biology involving both experimental and theoretical tools.

It is becoming increasingly common for biophysicists to apply the models and experimental techniques derived from physics, as well as mathematics and statistics, to larger systems such as tissues, organs, populations and ecosystems. Biophysical models are used extensively in the study of electrical conduction in single neurons, as well as neural circuit analysis in both tissue and whole brain.

History

Some of the earlier studies in biophysics were conducted in the 1840s by a group known as the Berlin school of physiologists. Among its members were pioneers such as Hermann von Helmholtz, Ernst Heinrich Weber, Carl F. W. Ludwig, and Johannes Peter Müller. Biophysics might even be seen as dating back to the studies of Luigi Galvani.

The popularity of the field rose when the book "What is life?" by Erwin Schrodinger was published. Since 1957 biophysicists have organized themselves into the Biophysical Society which now has about 7,000 members over the world.

Focus as a Subfield

Generally, biophysics does not have university-level departments of its own, but has presence as groups across departments within the fields of molecular biology, biochemistry, chemistry, computer science, mathematics, medicine, pharmacology, physiology, physics, and neuroscience. What follows is a list of examples of how each department applies its efforts toward the study of biophysics. This list is hardly all inclusive. Nor does each subject of study belong exclusively to any particular department. Each academic institution makes its own rules and there is much overlap between departments.

- *Biology and molecular biology* - Almost all forms of biophysics efforts are included in some biology department somewhere. To include some: gene regulation, single protein dynamics, bioenergetics, patch clamping, biomechanics.
- *Structural biology* - Ångstrom-resolution structures of proteins, nucleic acids, lipids, carbohydrates, and complexes thereof.
- Biochemistry and chemistry - biomolecular structure, siRNA, nucleic acid structure, structure-activity relationships.
- *Computer science* - Neural networks, biomolecular and drug databases.
- *Computational chemistry* - molecular dynamics simulation, molecular docking, quantum chemistry
- Bioinformatics - sequence alignment, structural alignment, protein structure prediction
- *Mathematics* - graph/network theory, population modelling, dynamical systems, phylogenetics.
- *Medicine and neuroscience* - tackling neural networks experimentally (brain slicing) as well as theoretically (computer models), membrane permitivity, gene therapy, understanding tumors.
- *Pharmacology and physiology* - channelomics, biomolecular interactions, cellular membranes, polyketides.
- *Physics* - negentropy, stochastic processes, covering dynamics.
- Quantum biophysics involves quantum information processing of coherent states, entanglement between coherent protons and transcriptase components, and replication of decohered isomers to yield time-dependent base substitutions. These studies imply applications in quantum computing.

Agronomy and agriculture

Many biophysical techniques are unique to this field. Research efforts in biophysics are often initiated by scientists who were traditional physicists, chemists, and biologists by training.

Nature and Subject of Biophysics

Biophysics is

- An academic discipline – branch of knowledge that is taught and researched at the college or university level. Disciplines are defined (in part), and recognized by the academic journals in which research is published, and the learned societies and academic departments or faculties to which their practitioners belong.

- A scientific field (a branch of science) – widely-recognized category of specialized expertise within science, and typically embodies its own terminology and nomenclature. Such a field will usually be represented by one or more scientific journals, where peer reviewed research is published.
    - A natural science – one that seeks to elucidate the rules that govern the natural world using empirical and scientific methods.
        - A biological science – concerned with the study of living organisms, including their structure, function, growth, evolution, distribution, and taxonomy.
        - A branch of physics – concerned with the study of matter and its motion through space and time, along with related concepts such as energy and force.
- An interdisciplinary field – field of science that overlaps with other sciences

Scope of Biophysics Research

- Biomolecular scale
    - Biomolecule
    - Biomolecular structure
- Organismal scale
    - Animal locomotion
    - Biomechanics
    - Biomineralization
    - Motility
- Environmental scale
    - Biophysical environment

Biophysics Research Overlaps with

- Agrophysics
- Biochemistry
    - Biophysical chemistry
- Bioengineering
- Biogeophysics

- Nanotechnology
- Systems biology

Branches of Biophysics

- Astrobiophysics – field of intersection between astrophysics and biophysics concerned with the influence of the astrophysical phenomena upon life on planet Earth or some other planet in general.
- Medical biophysics – uses physics to describe and control biological processes for diagnostic and therapeutic applications.
  - o Clinical biophysics – studies the process and effects of non-ionising physical energies utilised for diagnostic and therapeutic purposes.
- Membrane biophysics – study of biological membranes using physical, computational, mathematical, and biophysical methods.
- Molecular biophysics – interdisciplinary field that applies methods and concepts from physics, chemistry, engineering, mathematics and biology to understand biomolecular systems and explain biological function in terms of molecular structure, structural organization, and dynamic behaviour at various levels of complexity, from single molecules to supramolecular structures, viruses and small living systems.

Systems Biology

Systems biology (also known as Systeomics) is the computational and mathematical modelling of complex biological systems. An emerging engineering approach applied to biomedical and biological scientific research. Systems biology is a biology-based inter-disciplinary field of study that focuses on complex interactions within biological systems, using a holistic approach (holism instead of the more traditional reductionism) to biological and biomedical research. Particularly from year 2000 onwards, the concept has been used widely in the biosciences in a variety of contexts.

For example, the Human Genome Project is an example of applied systems thinking in biology which has led to new, collaborative ways of working on problems in the biological field of genetics. One of the outreaching aims of systems biology is to model and discover emergent properties, properties of cells, tissues and organisms functioning as a system whose theoretical description is only possible using techniques which fall under the remit of systems biology. These typically involve metabolic networks or cell signaling networks.

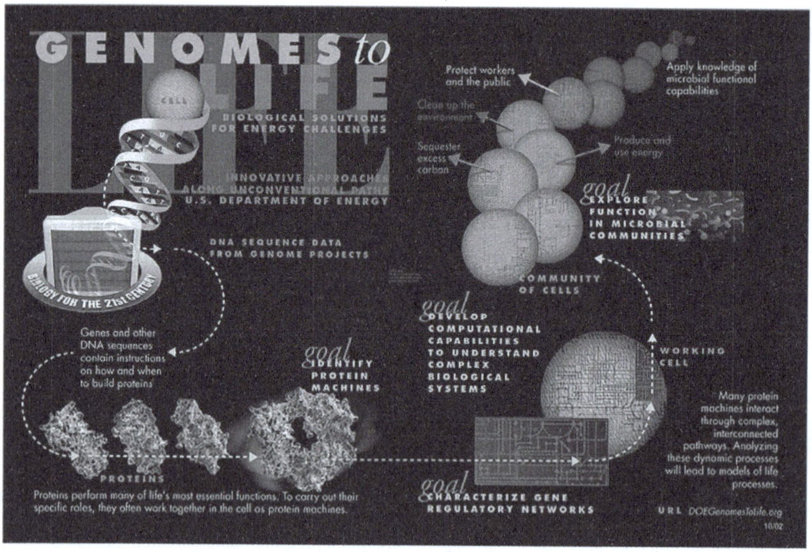

Figure: *An illustration of the systems approach to biology*

Systems biology can be considered from a number of different aspects:

- As a field of study, particularly, the study of the interactions between the components of biological systems, and how these interactions give rise to the function and behaviour of that system (for example, the enzymes and metabolites in a metabolic pathway).
- As a paradigm, usually defined in antithesis to the so-called reductionist paradigm (biological organisation), although fully consistent with the scientific method. The distinction between the two paradigms is referred to in these quotations:

  "The reductionist approach has successfully identified most of the components and many of the interactions but, unfortunately, offers no convincing concepts or methods to understand how system properties emerge...the pluralism of causes and effects in biological networks is better addressed by observing, through quantitative measures, multiple components simultaneously and by rigorous data integration with mathematical models" (Sauer et al.).

  "Systems biology...is about putting together rather than taking apart, integration rather than reduction. It

requires that we develop ways of thinking about integration that are as rigorous as our reductionist programmes, but different....It means changing our philosophy, in the full sense of the term" (Denis Noble).

- As a series of operational protocols used for performing research, namely a cycle composed of theory, analytic or computational modelling to propose specific testable hypotheses about a biological system, experimental validation, and then using the newly acquired quantitative description of cells or cell processes to refine the computational model or theory. Since the objective is a model of the interactions in a system, the experimental techniques that most suit systems biology are those that are system-wide and attempt to be as complete as possible. Therefore, transcriptomics, metabolomics, proteomics and high-throughput techniques are used to collect quantitative data for the construction and validation of models.
- As the application of dynamical systems theory to molecular biology. Indeed, the focus on the dynamics of the studied systems is the main conceptual difference between systems biology and bioinformatics.
- As a socioscientific phenomenon defined by the strategy of pursuing integration of complex data about the interactions in biological systems from diverse experimental sources using interdisciplinary tools and personnel.

This variety of viewpoints is illustrative of the fact that systems biology refers to a cluster of peripherally overlapping concepts rather than a single well-delineated field. However the term has widespread currency and popularity as of 2007, with chairs and institutes of systems biology proliferating worldwide.

The investigations are frequently combined with large-scale perturbation methods, including gene-based (RNAi, mis-expression of wild type and mutant genes) and chemical approaches using small molecule libraries. Robots and automated sensors enable such large-scale experimentation and data acquisition. These technologies are still emerging and many face problems that the larger the quantity of data produced, the lower the quality. A wide variety of quantitative scientists (computational biologists, statisticians, mathematicians, computer scientists, engineers, and physicists) are working to improve the quality of these approaches and to create, refine, and retest the models to accurately reflect observations.

The systems biology approach often involves the development of mechanistic models, such as the reconstruction of dynamic systems from the quantitative properties of their elementary building blocks.

For instance, a cellular network can be modelled mathematically using methods coming from chemical kinetics and control theory. Due to the large number of parameters, variables and constraints in cellular networks, numerical and computational techniques are often used (e.g., flux balance analysis).

Bioinformatics and Data Analysis

Other aspects of computer science, informatics, statistics are also used in systems biology. These include:

New forms of computational model, such as the use of process calculi to model biological processes (notable approaches include stochastic π-calculus, BioAmbients, Beta Binders, BioPEPA, and Brane calculus) and constraint-based modelling.

Integration of information from the literature, using techniques of information extraction and text mining.

Development of online databases and repositories for sharing data and models, approaches to database integration and software interoperability via loose coupling of software, websites and databases, or commercial suits.

Development of syntactically and semantically sound ways of representing biological models.

Network-based approaches for analyzing high dimensional genomic data sets. For example, weighted correlation network analysis is often used for identifying clusters (referred to as modules), modelling the relationship between clusters, calculating fuzzy measures of cluster (module) membership, identifying intramodular hubs, and for studying cluster preservation in other data sets.

Biophysical Techniques

Biophysical techniques – methods used for gaining information about biological systems on an atomic or molecular level. They overlap with methods from many other branches of science.

- Biophotonics – combination of biology and photonics, with photonics being the science and technology of generation, manipulation, and detection of photons, quantum units of light. Biophotonics can also be described as the "development and application of optical techniques, particularly imaging, to the study of biological molecules, cells and tissue". One of the main benefits of using optical techniques which make up biophotonics is that they preserve the integrity of the biological cells being examined.

- Biosensor and Bioelectronics –
- Calcium imaging –
- Calorimetry –
- Chromatography – various techniques from this field are used for the purification and analysis of biological molecules
- Circular Dichroism –
- Computational chemistry – use of numerical methods to probe the structure and dynamical equilibrium in biological systems.
- Cryobiology –
- Dual Polarisation Interferometry – analytical technique used to measure the real-time conformation and activity of a wide range of biomolecules and their interactions.
- Electrophysiology – studies electrical properties of cell membranes and provide functional data, often related to systematic changes in structure.
- Electron microscopy – used to gain high-resolution images of subcellular structures
- Fluorescence spectroscopy – for detecting structural rearrangements, as well as interactions of biomolecules.
- Force spectroscopy – probes the mechanical properties of individual molecules or macromolecular assemblies using small flexible cantilevers, focused laser light, or magnetic fields.
- Gel electrophoresis – determines the mass, the charge and the interactions of biological molecules
- Imaging – scientific imaging of biological materials, usually by some form of microscopy, or sometimes indirectly such as in x-ray crystallography or by computer imaging; at a wide range of magnifications to see macromolecules, cells, tissues, or organisms
- Isothermal Titration Calorimetry (ITC) – measures the heat effects caused by interactions
- Mass spectrometry – technique that gives the molecular mass with great accuracy.

- Microscale Thermophoresis (MST) – method to measure binding affinities, enzymatic activities, changes in molecule conformation and changes in size, charge or hydration entropy.
- Microscopy – used in many ways, for example, to enabled the use of laser instruments for scanning and transmission.
  - o Atomic force microscopy –
- Neuroimaging –
- Neutron spin echo spectroscopy –
- Optical tweezers and Magnetic tweezers – allow for the manipulation of single molecules, providing information about DNA and its interaction with proteins and molecular motors, such as Helicase and RNA polymerase.
- Patch clamping –
- NMR spectroscopy – provides information about the exact structure of biological molecules, as well as on dynamics
- Single molecule spectroscopy – general term applied to a class of techniques that are sensitive enough to detect single molecules and often incorporates fluorescence detection.
- Small angle X-ray scattering (SAXS) – technique that gives a rough low resolution molecular structure.
- Spectrophotometry – measurement of the transmission of light through different solutions or substances at different wavelengths of light.
  - o Colorimetry –
- Spectroscopy and Circular dichroism – method for detecting chiral groups in molecules, especially to determine the secondary structure of proteins
  - o Nuclear Magnetic Resonance Spectroscopy –
- Ultracentrifugation – gives information on the shape and mass of molecules
- X-ray crystallography – method to determine the exact structure of molecules with atomic resolution

Applications

A biophysical profile (BPP) is a prenatal ultrasound evaluation of fetal well-being involving a scoring system, with the score being termed Manning's score. It is often done when a non-stress test (NST) is non reactive, or for other obstetrical indications.

The "modified biophysical profile" consists of the NST and amniotic fluid index only.

The BPP has 5 components: 4 ultrasound (US) assessments and a NST. The NST evaluates fetal heart rate and response to fetal movement. The five discrete biophysical variables:

- Fetal movement
- Fetal tone
- Fetal breathing
- Amniotic fluid volume
- Fetal Heart Rate

Molecular Structure of Biological Systems

A biological system is a complex network of biologically relevant entities. As biological organization spans several scales, examples of biological systems are populations of organisms, or on the organ- and tissue scale in mammals and other animals, the circulatory system, the respiratory system, the nervous system, etc.

On the micro- to the nanoscopic scale, examples of biological systems are cells, organelles, macromolecular complexes and regulatory pathways.

A biological system is not to be confused with a living system, which is commonly referred to as life. For further information see e.g. definition of life or synthetic biology.

Biomolecule

A biomolecule is any molecule that is produced by a living organism, including large macromolecules such as proteins, polysaccharides, lipids, and nucleic acids, as well as small molecules such as primary metabolites, secondary metabolites, and natural products. A more general name for this class of molecules is biogenic substances.

Types of Biomolecules

A diverse range of biomolecules exist, including:

Small molecules:

Lipids, fatty acids, glycolipids, sterols, glycerolipids

Vitamins

Hormones, neurotransmitters

Metabolites

Monomers, oligomers and polymers:

Biomonomers	*Bio-oligomers*	*Biopolymers*	*Polymerization process*	*Covalent bond name between monomers*
Amino acids	Oligopeptides	Polypeptides, proteins (hemoglobin...)	Polycondensation	Peptide bond
Monosaccharides	Oligosaccharides	Polysaccharides (cellulose...)	Polycondensation	Glycosidic bond
Isoprene	Terpenes	Polyterpenes: cis-1,4-polyisoprene natural rubber and trans-1,4-polyisoprene gutta-percha	Polyaddition	
Nucleotides	Oligonucleotides	Polynucleotides, nucleic acids (DNA, RNA)		Phosphodiester bond

Nucleosides and Nucleotides

Nucleosides are molecules formed by attaching a nucleobase to a ribose or deoxyribose ring. Examples of these include cytidine (C), uridine (U), adenosine (A), guanosine (G), thymidine (T) and inosine (I).

Nucleosides can be phosphorylated by specific kinases in the cell, producing nucleotides. Both DNA and RNA are polymers, consisting of long, linear molecules assembled by polymerase enzymes from repeating structural units, or monomerss, of mononucleotides. DNA uses the deoxynucleotides C, G, A, and T, while RNA uses the ribonucleotides (which have an extra hydroxyl(OH) group on the pentose ring) C, G, A, and U. Modified bases are fairly common (such as with methyl groups on the base ring), as found in ribosomal RNA or transfer RNAs or for discriminating the new from old strands of DNA after replication.

Each nucleotide is made of an acyclic nitrogenous base, a pentose and one to three phosphate groups. They contain carbon, nitrogen, oxygen, hydrogen and phosphorus. They serve as sources of chemical energy (adenosine triphosphate and guanosine triphosphate), participate in cellular signaling (cyclic guanosine monophosphate and cyclic adenosine monophosphate), and are incorporated into important cofactors of enzymatic reactions (coenzyme A, flavin adenine dinucleotide, flavin mononucleotide, and nicotinamide adenine dinucleotide phosphate).

DNA and RNA Structure

DNA structure is dominated by the well-known double helix formed by Watson-Crick base-pairing of C with G and A with T. This is known as B-form DNA, and is overwhelmingly the most favourable and common state of DNA; its highly specific and stable base-pairing is the basis of reliable genetic information storage.

DNA can sometimes occur as single strands (often needing to be stabilized by single-strand binding proteins) or as A-form or Z-form helices, and occasionally in more complex 3D structures such as the crossover at Holliday junctions during DNA replication.

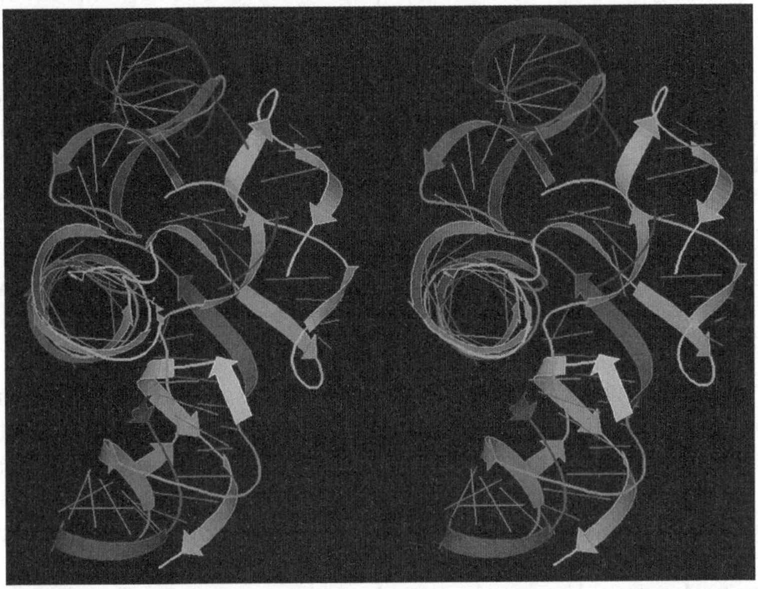

Figure: *Stereo 3D image of a group I intron ribozyme (PDB file 1Y0Q); gray lines show base pairs; ribbon arrows show double-helix regions, blue to red from 5' to 3' end; white ribbon is an RNA product.*

RNA, in contrast, forms large and complex 3D tertiary structures reminiscent of proteins, as well as the loose single strands with locally folded regions that constitute messenger RNA molecules. Those RNA structures contain many stretches of A-form double helix, connected into definite 3D arrangements by single-stranded loops, bulges, and junctions. Examples are tRNA, ribosomes, ribozymes, and riboswitches.

These complex structures are facilitated by the fact that RNA backbone has less local flexibility than DNA but a large set of distinct conformations, apparently because of both positive and negative interactions of the extra OH on the ribose. Structured RNA molecules can do highly specific binding of other molecules and can themselves be recognized specifically; in addition, they can perform enzymatic catalysis (when they are known as "ribozymes", as initially discovered by Tom Cech and colleagues.

Saccharides

Monosaccharides are the simplest form of carbohydrates with only one simple sugar. They essentially contain an aldehyde or ketone group in their structure. The presence of an aldehyde group in a monosaccharide is indicated by the prefix *aldo-*. Similarly, a ketone group is denoted by the prefix *keto-*. Examples of monosaccharides are the hexoses glucose, fructose, and galactose and pentoses, ribose, and deoxyribose Consumed fructose and glucose have different rates of gastric emptying, are differentially absorbed and have different metabolic fates, providing multiple opportunities for 2 different saccharides to differentially affect food intake. Most saccharides eventually provide fuel for cellular respiration.

Disaccharides are formed when two monosaccharides, or two single simple sugars, form a bond with removal of water. They can be hydrolyzed to yield their saccharin building blocks by boiling with dilute acid or reacting them with appropriate enzymes. Examples of disaccharides include sucrose, maltose, and lactose.

Polysaccharides are polymerized monosaccharides, or complex carbohydrates. They have multiple simple sugars. Examples are starch, cellulose, and glycogen. They are generally large and often have a complex branched connectivity. Because of their size, polysaccharides are not water-soluble, but their many hydroxy groups become hydrated individually when exposed to water, and some polysaccharides form thick colloidal dispersions when heated in water. Shorter polysaccharides, with 3 - 10 monomers, are called oligosaccharides. A fluorescent indicator-displacement molecular imprinting sensor was developed for discriminating saccharides. It successfully discriminated three brands of orange juice beverage. The change in fluorescence intensity of the sensing films resulting is directly related to the saccharide concentration.

Lignin

Lignin is a complex polyphenolic macromolecule composed mainly of beta-O4-aryl linkages. After cellulose, lignin is the second most abundant biopolymer and is one of the primary structural components of most plants. It contains subunits derived from *p*-coumaryl alcohol, coniferyl alcohol, and sinapyl alcohol and is unusual among biomolecules in that it is racemic. The lack of optical activity is due to the polymerization of lignin which occurs via free radical coupling reactions in which there is no preference for either configuration at a chiral centre.

Lipids

Lipids (oleaginous) are chiefly fatty acid esters, and are the basic building blocks of biological membranes. Another biological role is energy storage (e.g., triglycerides). Most lipids consist of a polar or hydrophilic head (typically glycerol) and one to three nonpolar or hydrophobic fatty acid tails, and therefore they are amphiphilic. Fatty acids consist of unbranched chains of carbon atoms that are connected by single bonds alone (saturated fatty acids) or by both single and double bonds (unsaturated fatty acids). The chains are usually 14-24 carbon groups long, but it is always an even number.

For lipids present in biological membranes, the hydrophilic head is from one of three classes:

- Glycolipids, whose heads contain an oligosaccharide with 1-15 saccharide residues.
- Phospholipids, whose heads contain a positively charged group that is linked to the tail by a negatively charged phosphate group.
- Sterols, whose heads contain a planar steroid ring, for example, cholesterol.

Other lipids include prostaglandins and leukotrienes which are both 20-carbon fatty acyl units synthesized from arachidonic acid. They are also known as fatty acids

Amino Acids

Amino acids contain both amino and carboxylic acid functional groups. (In biochemistry, the term amino acid is used when referring to those amino acids in which the amino and carboxylate functionalities are attached to the same carbon, plus proline which is not actually an amino acid).

Modified amino acids are sometimes observed in proteins; this is usually the result of enzymatic modification after translation (protein synthesis). For example, phosphorylation of serine by kinases and dephosphorylation by phosphatases is an important control mechanism in the cell cycle. Only two amino acids other than the standard twenty are known to be incorporated into proteins during translation, in certain organisms:

- Selenocysteine is incorporated into some proteins at a UGA codon, which is normally a stop codon.
- Pyrrolysine is incorporated into some proteins at a UAG codon. For instance, in some methanogens in enzymes that are used to produce methane.

Besides those used in protein synthesis, other biologically important amino acids include carnitine (used in lipid transport within a cell), ornithine, GABA and taurine.

Protein Structure

The particular series of amino acids that form a protein is known as that protein's primary structure. This sequence is determined by the genetic makeup of the individual. It specifies the order of side-chain groups along the linear polypeptide "backbone".

Proteins have two types of well-classified, frequently occurring elements of local structure defined by a particular pattern of hydrogen bonds along the backbone: alpha helix and beta sheet. Their number and arrangement is called the secondary structure of the protein. Alpha helices are regular spirals stabilized by hydrogen bonds between the backbone CO group (carbonyl) of one amino acid residue and the backbone NH group (amide) of the i+4 residue.

The spiral has about 3.6 amino acids per turn, and the amino acid side chains stick out from the cylinder of the helix. Beta pleated sheets are formed by backbone hydrogen bonds between individual beta strands each of which is in an "extended", or fully stretched-out, conformation. The strands may lie parallel or antiparallel to each other, and the side-chain direction alternates above and below the sheet. Hemoglobin contains only helices, natural silk is formed of beta pleated sheets, and many enzymes have a pattern of alternating helices and beta-strands. The secondary-structure elements are connected by "loop" or "coil" regions of non-repetitive conformation, which are sometimes quite mobile or disordered but usually adopt a well-defined, stable arrangement.

The overall, compact, 3D structure of a protein is termed its tertiary structure or its "fold". It is formed as result of various attractive forces like hydrogen bonding, disulfide bridges, hydrophobic interactions, hydrophilic interactions, van der Waals force etc.

When two or more polypeptide chains (either of identical or of different sequence) cluster to form a protein, quaternary structure of protein is formed. Quaternary structure is an attribute of polymeric (same-sequence chains) or heteromeric (different-sequence chains) proteins like hemoglobin, which consists of two "alpha" and two "beta" polypeptide chains.

Apoenzymes

An apoenzyme (or, generally, an apoprotein) is the protein without any small-molecule cofactors, substrates, or inhibitors bound. It is often important as an inactive storage, transport, or secretory form of a protein. This is required, for instance, to protect the secretory cell from the activity of that protein.

Apoenzymes becomes active enzymes on addition of a cofactor. Cofactors can be either inorganic (e.g., metal ions and iron-sulfur clusters) or organic compounds, (e.g., flavin and heme). Organic cofactors can be either prosthetic groups, which are tightly bound to an enzyme, or coenzymes, which are released from the enzyme's active site during the reaction.

Isoenzymes

Isoenzymes, or isozymes, are multiple forms of an enzyme, with slightly different protein sequence and closely similar but usually not identical functions. They are either products of different genes, or else different products of alternative splicing. They may either be produced in different organs or cell types to perform the same function, or several isoenzymes may be produced in the same cell type under differential regulation to suit the needs of changing development or environment. LDH (lactate dehydrogenase) has multiple isozymes, while fetal hemoglobin is an example of a developmentally regulated isoform of a non-enzymatic protein. The relative levels of isoenzymes in blood can be used to diagnose problems in the organ of secretion.

Vitamins

A vitamin is a compound that is generally not synthesized by a given organism but is nonetheless vital to its survival or health (for example coenzymes). These compounds must be absorbed, or eaten, but typically only in trace quantities. When originally proposed by Casimir Funk, a Polish biochemist, he believed them to all be basic and therefore named them vital amines. The "al" was later dropped to form the word vitamines.

Energy Balance

In biology, energy homeostasis, or energy balance, are an aspect of bioenergetics concerning the energy flow through living systems, by metabolism. Energy homeostasis involves the human body using chemical and neural signals to adjust the amount of energy flows; and to regulate caloric intake by signaling the brain to regulate the sensation of hunger. Fifty percent of the energy from glucose metabolism is immediately converted to heat.

Energy balance, through biosynthetic reactions, can be measured with the following equation:

Energy intake (food) = Energy expended (heat + work) + Energy stored.

The first law of thermodynamics states that energy can be neither created nor destroyed. But energy can be converted from one form of energy to another. So when a calorie of food eaten enters a body, ultimately 100% of that calorie will be converted to heat, resulting in three particular short-term effects: a portion of that calorie is either stored as fat, transferred to the body's cells as chemical energy, by Adenosine triphosphate (ATP), a coenzyme and used for mechanical work or chemical synthesis, or immediately dissipated through heat.

Energy

Intake

Energy intake is calories consumed as food, which is determined by hunger (hypothalamus) and choice (cerebral cortex). Hunger is determined by the impact of the hormones leptin and ghrelin on the hypothalamus.

Expenditure

Energy expenditure is mainly a sum of internal heat produced and external work. The internal heat produced is, in turn, mainly a sum of basal metabolic rate (BMR) and the thermic effect of food. External work may be estimated by measuring the physical activity level (PAL).

Imbalance

Positive Balance: A positive balance is a result of *energy intake* being *higher* than what is consumed in external work and other bodily means of energy expenditure.

The main preventable causes are:

- Overeating, resulting in increased energy intake
- Sedentary lifestyle, resulting in decreased energy expenditure through external work

A positive balance results in energy being stored as fat and/or muscle, causing weight gain. In time, overweight and obesity may develop, with resultant complications.

Negative Balance

A negative balance is a result of energy intake being less than what is consumed in external work and other bodily means of energy expenditure.

The main cause is *undereating* due to a medical condition such as decreased appetite, anorexia nervosa, digestive disease, or due to some circumstance such as fasting or lack of access to food

Requirement

Normal energy requirement, and therefore normal energy intake, depends mainly on age, sex and physical activity level (PAL). The Food and Agriculture Organization (FAO) of the United Nations has compiled a detailed report on human energy requirements: Human energy requirements (Rome, 17–24 October 2001) An older but commonly used and fairly accurate method is the Harris-Benedict equation.

Yet, there are currently ongoing studies to show if calorie restriction to below normal values have beneficial effects, and even though they are showing positive indications in primates it is still not certain if calorie restriction has a positive effect on longevity for primates and humans.

Calorie restriction may be viewed as attaining energy balance at a lower intake and expenditure, and is, in this sense, not generally an energy imbalance, except for an initial imbalance where decreased expenditure hasn't yet matched the decreased intake.

Physical Factors of the Environment

The biophysical environment is the biotic and abiotic surrounding of an organism or population, and consequently includes the factors that have an influence in their survival, development and evolution. The term environment can refer to different concepts, but is often used as a short form for the *biophysical environment.*

This practice is common, for instance, among governments which entitle agencies dealing with the biophysical environment with denominations such as Environment Agency.

Whereas the expression "*the environment*" is often used to refer to the global environment, usually in relation to humanity, the number of biophysical environments is countless, given that it is always possible to consider an additional living organism that has its own environment.

Constituents

The biophysical environment can vary in scale from microscopic to global in extent. It can also be subdivided according to its attributes. Examples include the marine environment, the atmospheric environment and the terrestrial environment.

Life/Environment Interaction

All life that has survived must have adapted to conditions of its environment. Temperature, light, humidity, soil nutrients, etc., all influence any species, within any environment. However life in turn modifies, in various forms, its conditions. Some long term modifications along the history of our planet have been significant, such as the incorporation of oxygen to the atmosphere. This process consisted in the breakdown of carbon dioxide by anaerobic microorganisms that used the carbon in their metabolism and released the oxygen to the atmosphere. This led to the existence of oxygen-based plant and animal life, the great oxygenation event. Other interactions are more immediate and simple, such as the smoothing effect that forests have on the temperature cycle, compared to neighbouring unforested areas.

Definitions

Environmental science is the study of the interactions within the biophysical environment. Part of this scientific discipline is the investigation of the effect of human activity on the environment. Ecology, a sub-discipline of biology and a part of environmental sciences, is often mistaken as a study of human induced effects on the environment. Environmental studies is a broader academic discipline that is the systematic study of interaction of humans with their environment. It is a broad field of study that includes the natural environment, built environments and social environments.

Environmentalism is a broad social and philosophical movement that, in a large part, seeks to minimise and compensate the negative effect of human activity on the biophysical environment. The issues of concern for environmentalists usually relate to the natural environment with the more important ones being climate change, species extinction, pollution, and old growth forest loss.

Enzyme Kinetics of Biological Systems

Enzyme kinetics is the study of the chemical reactions that are catalysed by enzymes. In enzyme kinetics, the reaction rate is measured and the effects of varying the conditions of the reaction are investigated. Studying an enzyme's kinetics in this way can reveal the catalytic mechanism of this enzyme, its role in metabolism, how its activity is controlled, and how a drug or an agonist might inhibit the enzyme.

Enzymes are usually protein molecules that manipulate other molecules — the enzymes' substrates. These target molecules bind to an enzyme's active site and are transformed into products through a series of steps known as the enzymatic mechanism. These mechanisms can be divided into single-substrate and multiple-substrate mechanisms. Kinetic studies on enzymes that only bind one substrate, such as triosephosphate isomerase, aim to measure the affinity with which the enzyme binds this substrate and the turnover rate. Some other examples of enzymes are phosphofructokinase and hexokinase, both of which are important for cellular respiration (glycolysis).

Figure: *Dihydrofolate reductase from E. coli with its two substrates dihydrofolate (right) and NADPH (left), bound in the active site. The protein is shown as a ribbon diagram, with alpha helices in red, beta sheets in yellow and loops in blue. Generated from 7DFR.*

When enzymes bind multiple substrates, such as dihydrofolate reductase (shown right), enzyme kinetics can also show the sequence in which these substrates bind and the sequence in which products are released. An example of enzymes that bind a single substrate and release multiple products are proteases, which cleave one protein substrate into two polypeptide products. Others join two substrates together, such as DNA polymerase linking a nucleotide to DNA. Although these mechanisms are often a complex series of steps, there is typically one *rate-determining step* that determines the overall kinetics. This rate-determining step may be a chemical reaction or a conformational change of the enzyme or substrates, such as those involved in the release of product(s) from the enzyme.

Knowledge of the enzyme's structure is helpful in interpreting kinetic data. For example, the structure can suggest how substrates and products bind during catalysis; what changes occur during the reaction; and even the role of particular amino acid residues in the mechanism. Some enzymes change shape significantly during the mechanism; in such cases, it is helpful to determine the enzyme structure with and without bound substrate analogues that do not undergo the enzymatic reaction.

Not all biological catalysts are protein enzymes; RNA-based catalysts such as ribozymes and ribosomes are essential to many cellular functions, such as RNA splicing and translation. The main difference between ribozymes and enzymes is that RNA catalysts are composed of nucleotides, whereas enzymes are composed of amino acids. Ribozymes also perform a more limited set of reactions, although their reaction mechanisms and kinetics can be analysed and classified by the same methods.

General Principles

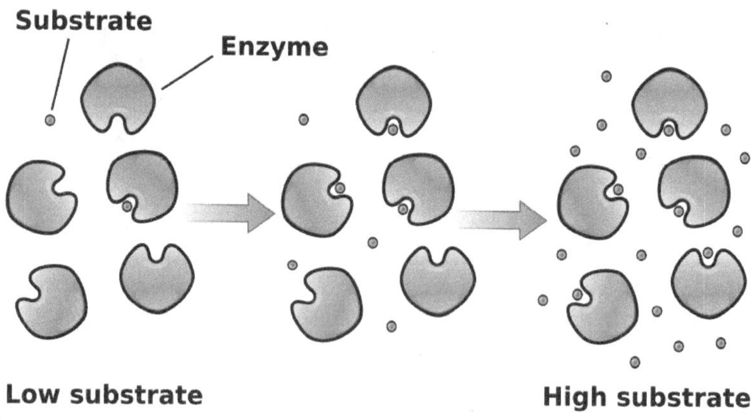

Figure: *As larger amounts of substrate are added to a reaction, the available enzyme binding sites become filled to the limit of* V_{max}. *Beyond this limit the enzyme is saturated with substrate and the reaction rate ceases to increase.*

The reaction catalysed by an enzyme uses exactly the same reactants and produces exactly the same products as the uncatalysed reaction. Like other catalysts, enzymes do not alter the position of equilibrium between substrates and products. However, unlike uncatalysed chemical reactions, enzyme-catalysed reactions display saturation kinetics. For a given enzyme concentration and for relatively low substrate concentrations, the reaction rate increases linearly with substrate concentration; the enzyme molecules are largely free to catalyse the reaction, and increasing substrate concentration means an increasing rate at which the enzyme and substrate molecules encounter one another.

However, at relatively high substrate concentrations, the reaction rate asymptotically approaches the theoretical maximum; the enzyme active sites are almost all occupied and the reaction rate is determined by the intrinsic turnover rate of the enzyme. The substrate concentration midway between these two limiting cases is denoted by K_M.

The two most important kinetic properties of an enzyme are how quickly the enzyme becomes saturated with a particular substrate, and the maximum rate it can achieve. Knowing these properties suggests what an enzyme might do in the cell and can show how the enzyme will respond to changes in these conditions.

Enzyme Assays

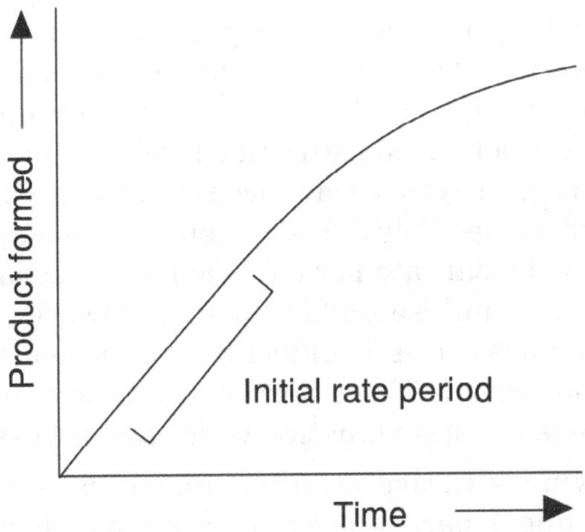

***Figure**: Progress curve for an enzyme reaction. The slope in the initial rate period is the initial rate of reaction* v. *The Michaelis–Menten equation describes how this slope varies with the concentration of substrate.*

Enzyme assays are laboratory procedures that measure the rate of enzyme reactions. Because enzymes are not consumed by the reactions they catalyse, enzyme assays usually follow changes in the concentration of either substrates or products to measure the rate of reaction. There are many methods of measurement. Spectrophotometric assays observe change in the absorbance of light between products and reactants; radiometric assays involve the incorporation or release of radioactivity to measure the amount of product made over time. Spectrophotometric assays are most convenient since they allow the rate of the reaction to be measured continuously. Although radiometric assays require the removal and counting of samples (i.e., they are discontinuous assays) they are usually extremely sensitive and can measure very low levels of enzyme activity. An analogous approach is to use mass spectrometry to monitor the incorporation or release of stable isotopes as substrate is converted into product.

The most sensitive enzyme assays use lasers focused through a microscope to observe changes in single enzyme molecules as they catalyse their reactions. These measurements either use changes in the fluorescence of cofactors during an enzyme's reaction mechanism, or of fluorescent dyes added onto specific sites of the protein to report movements that occur during catalysis. These studies are providing a new view of the kinetics and dynamics of single enzymes, as opposed to traditional enzyme kinetics, which observes the average behaviour of populations of millions of enzyme molecules.

An example progress curve for an enzyme assay is shown above. The enzyme produces product at an initial rate that is approximately linear for a short period after the start of the reaction. As the reaction proceeds and substrate is consumed, the rate continuously slows (so long as substrate is not still at saturating levels). To measure the initial (and maximal) rate, enzyme assays are typically carried out while the reaction has progressed only a few percent towards total completion. The length of the initial rate period depends on the assay conditions and can range from milliseconds to hours. However, equipment for rapidly mixing liquids allows fast kinetic measurements on initial rates of less than one second. These very rapid assays are essential for measuring pre-steady-state kinetics, which are discussed below.

Most enzyme kinetics studies concentrate on this initial, approximately linear part of enzyme reactions. However, it is also possible to measure the complete reaction curve and fit this data to a non-linear rate equation. This way of measuring enzyme reactions is called progress-curve analysis. This approach is useful as an alternative to rapid kinetics when the initial rate is too fast to measure accurately.

Single-Substrate Reactions

Enzymes with single-substrate mechanisms include isomerases such as triosephosphateisomerase or bisphosphoglycerate mutase, intramolecular lyases such as adenylate cyclase and the hammerhead ribozyme, an RNA lyase. However, some enzymes that only have a single substrate do not fall into this category of mechanisms.

Catalase is an example of this, as the enzyme reacts with a first molecule of hydrogen peroxide substrate, becomes oxidised and is then reduced by a second molecule of substrate.

Although a single substrate is involved, the existence of a modified enzyme intermediate means that the mechanism of catalase is actually a ping–pong mechanism, a type of mechanism that is discussed in the *Multi-substrate reactions* section below.

Michaelis–Menten Kinetics

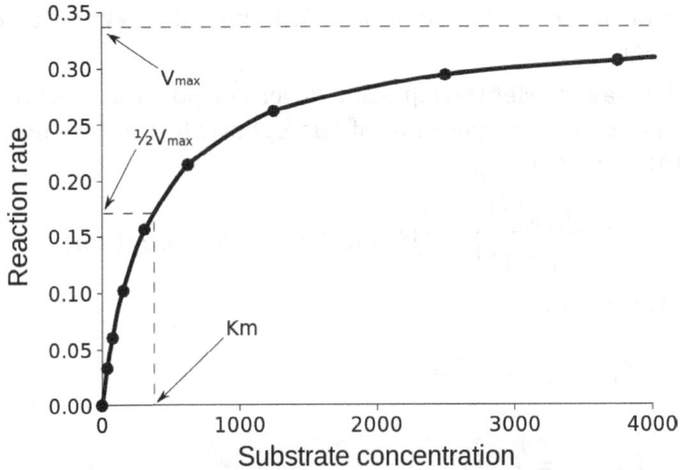

Figure: *Saturation curve for an enzyme showing the relation between the concentration of substrate and rate.*

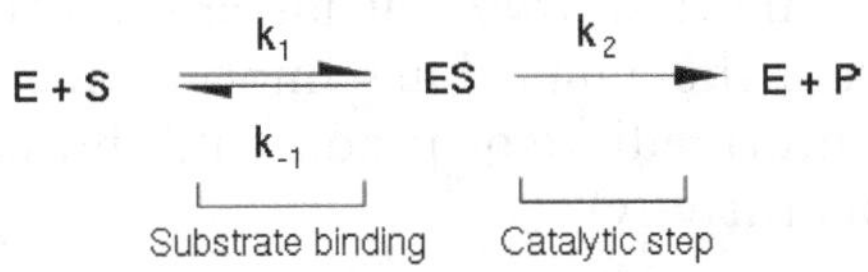

Figure: *Single-substrate mechanism for an enzyme reaction.* k_1, k_{-1} *and* k_2 *are the rate constants for the individual steps.*

As enzyme-catalysed reactions are saturable, their rate of catalysis does not show a linear response to increasing substrate. If the initial rate of the reaction is measured over a range of substrate concentrations (denoted as [S]), the reaction rate (v) increases as [S] increases, as shown on the right. However, as [S] gets higher, the enzyme becomes saturated with substrate and the rate reaches V_{max}, the enzyme's maximum rate.

The Michaelis–Menten kinetic model of a single-substrate reaction is shown on the right. There is an initial bimolecular reaction between the enzyme E and substrate S to form the enzyme–substrate complex ES. Although the enzymatic mechanism for the unimolecular reaction $ES \xrightarrow{k_{cat}} E + P$ can be quite complex, there is typically one rate-determining enzymatic step that allows this reaction to be modelled as a single catalytic step with an apparent unimolecular rate constant k_{cat}. If the reaction path proceeds over one or several intermediates, k_{cat} will be a function of several elementary rate constants, whereas in the simplest case of a single elementary reaction (e.g. no intermediates) it will be identical to the elementary unimolecular rate constant k_2. The apparent unimolecular rate constant k_{cat} is also called *turnover number* and denotes the maximum number of enzymatic reactions catalysed per second.

The Michaelis–Menten equation describes how the (initial) reaction rate v_0 depends on the position of the substrate-binding equilibrium and the rate constant k_2.

$$v_0 = \frac{V_{\max}[S]}{K_M + [S]} \qquad \textit{(Michaelis–Menten equation)}$$

with the constants

$$K_M \overset{\text{def}}{=} \frac{k_2 + k_{-1}}{k_1} \approx K_D$$

$$V_{\max} \overset{\text{def}}{=} k_{cat}[E]_{tot}$$

This Michaelis–Menten equation is the basis for most single-substrate enzyme kinetics. Two crucial assumptions underlie this equation (apart from the general assumption about the mechanism only involving no intermediate or product inhibition, and there is no allostericity or cooperativity).

The first assumption is the so-called quasi-steady-state assumption (or pseudo-steady-state hypothesis), namely that the concentration of the substrate-bound enzyme (and hence also the unbound enzyme) changes much more slowly than those of the product and substrate and thus the change over time of the complex can be set to zero $d[ES]/dt \overset{!}{=} 0$. The second assumption is that the total enzyme concentration does not change over time, thus $[E]_{tot} = [E] + [ES] \overset{!}{=} \text{const}$. A complete derivation can be found here.

The Michaelis constant K_M is experimentally defined as the concentration at which the rate of the enzyme reaction is half V_{max}, which can be verified by substituting [S] = K_m into the Michaelis–Menten equation and can also be seen graphically. If the rate-determining enzymatic step is slow compared to substrate dissociation ($k_2 \ll k_{-1}$), the Michaelis constant K_M is roughly the dissociation constant K_D of the ES complex. If $[S]$ is small compared to K_M then the term $[S]/(K_M+[S]) \approx [S]/K_M$ and also very little ES complex is formed, thus $[E]_0 \approx [E]$. Therefore, the rate of product formation is

$$v_0 \approx \frac{k_{cat}}{K_M}[E][S] \qquad \text{if } [S] \ll K_M$$

Thus the product formation rate depends on the enzyme concentration as well as on the substrate concentration, the equation resembles a bimolecular reaction with a corresponding pseudo-second order rate constant k_2/K_M. This constant is a measure of catalytic efficiency. The most efficient enzymes reach a k_2/K_M in the range of $10^8 - 10^{10}$ M^{-1} s^{-1}. These enzymes are so efficient they effectively catalyse a reaction each time they encounter a substrate molecule and have thus reached an upper theoretical limit for efficiency (diffusion limit); these enzymes have often been termed *perfect enzymes*.

Direct Use of the Michaelis–Menten Equation for Time Course Kinetic Analysis

The observed velocities predicted by the Michaelis–Menten equation can be used to directly model the time course disappearance of substrate and the production of product through incorporation of the Michaelis–Menten equation into the equation for first order chemical kinetics.

This can only be achieved however if one recognises the problem associated with the use of Euler's number in the description of first order chemical kinetics. i.e. e^{-k} is a split constant that introduces a systematic error into calculations and can be rewritten as a single constant which represents the remaining substrate after each time period.

$$[S] = [S]_0(1-k)^t$$

$$[S] = [S]_0(1-v/[S]_0)^t$$

$$[S] = [S]_0(1-(V_{max}[S]_0/(K_M+[S]_0)/[S]_0))^t$$

In 1983 Stuart Beal (and also independently Santiago Schnell and Claudio Mendoza in 1997) derived a closed form solution for the time course kinetics analysis of the Michaelis-Menten mechanism. The solution, known as the Schnell-Mendoza equation, has the form:

$$\frac{[S]}{K_M} = W[F(t)]$$

where W[] is the Lambert-W function. and where F(t) is

$$F(t) = \frac{[S]_0}{K_M} \exp\left(\frac{[S]_0}{K_M} - \frac{V_{max}}{K_M} t\right)$$

This equation is encompassed by the equation below, obtained by Berberan-Santos (MATCH Commun. Math. Comput. Chem. 63 (2010) 283), which is also valid when the initial substrate concentration is close to that of enzyme,

$$\frac{[S]}{K_M} = W[F(t)] - \frac{V_{max}}{k_{cat} K_M} \frac{W[F(t)]}{1 + W[F(t)]}$$

where W[] is again the Lambert-W function.

Linear Plots of the Michaelis–Menten Equation

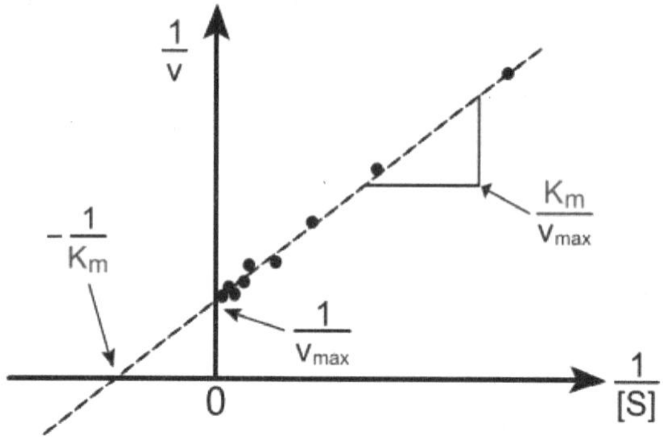

Figure: *Lineweaver–Burk or double-reciprocal plot of kinetic data, showing the significance of the axis intercepts and gradient.*

The plot of v versus [S] above is not linear; although initially linear at low [S], it bends over to saturate at high [S]. Before the modern era of nonlinear curve-fitting on computers, this nonlinearity could make it difficult to estimate K_M and V_{max} accurately. Therefore, several researchers developed linearisations of the Michaelis–Menten equation, such as the Lineweaver–Burk plot, the Eadie–Hofstee diagram and the Hanes–Woolf plot. All of these linear representations can be useful for visualising data, but none should be used to determine kinetic parameters, as computer software is readily available that allows for more accurate determination by nonlinear regression methods.

The Lineweaver–Burk plot or double reciprocal plot is a common way of illustrating kinetic data. This is produced by taking the reciprocal of both sides of the Michaelis–Menten equation. As shown on the right, this is a linear form of the Michaelis–Menten equation and produces a straight line with the equation $y = mx + c$ with a y-intercept equivalent to $1/V_{max}$ and an x-intercept of the graph representing "$1/K_M$.

$$\frac{1}{v} = \frac{K_M}{V_{max}[S]} + \frac{1}{V_{max}}$$

Naturally, no experimental values can be taken at negative 1/[S]; the lower limiting value 1/[S] = 0 (the y-intercept) corresponds to an infinite substrate concentration, where $1/v=1/V_{max}$ as shown at the right; thus, the x-intercept is an extrapolation of the experimental data taken at positive concentrations. More generally, the Lineweaver–Burk plot skews the importance of measurements taken at low substrate concentrations and, thus, can yield inaccurate estimates of V_{max} and K_M. A more accurate linear plotting method is the Eadie-Hofstee plot. In this case, v is plotted against v/[S]. In the third common linear representation, the Hanes-Woolf plot, [S]/v is plotted against [S]. In general, data normalisation can help diminish the amount of experimental work and can increase the reliability of the output, and is suitable for both graphical and numerical analysis.

Practical Significance of Kinetic Constants

The study of enzyme kinetics is important for two basic reasons. Firstly, it helps explain how enzymes work, and secondly, it helps predict how enzymes behave in living organisms. The kinetic constants defined above, K_M and V_{max}, are critical to attempts to understand how enzymes work together to control metabolism.

Making these predictions is not trivial, even for simple systems. For example, oxaloacetate is formed by malate dehydrogenase within the mitochondrion. Oxaloacetate can then be consumed by citrate synthase, phosphoenolpyruvate carboxykinase or aspartate aminotransferase, feeding into the citric acid cycle, gluconeogenesis or aspartic acid biosynthesis, respectively. Being able to predict how much oxaloacetate goes into which pathway requires knowledge of the concentration of oxaloacetate as well as the concentration and kinetics of each of these enzymes. This aim of predicting the behaviour of metabolic pathways reaches its most complex expression in the synthesis of huge amounts of kinetic and gene expression data into mathematical models of entire organisms. Alternatively, one useful simplification of the metabolic modelling problem is to ignore the underlying enzyme kinetics and only rely on information about the reaction network's stoichiometry, a technique called flux balance analysis.

Michaelis–Menten Kinetics with Intermediate

One could also consider the less simple case

$$E + S \underset{k_{-1}}{\overset{k_1}{\rightleftarrows}} ES \xrightarrow{k_2} EI \xrightarrow{k_3} E + P$$

where a complex with the enzyme and an intermediate exists and the intermediate is converted into product in a second step. In this case we have a very similar equation

$$v_0 = k_{cat} \frac{[S][E]_0}{K_M' + [S]}$$

but the constants are different

$$K_M' \overset{\text{def}}{=} \frac{k_3}{k_2 + k_3} K_M = \frac{k_3}{k_2 + k_3} \cdot \frac{k_2 + k_{-1}}{k_1}$$

$$k_{cat} \overset{\text{def}}{=} \frac{k_3 k_2}{k_2 + k_3}$$

We see that for the limiting case $k_3 \gg k_2$, thus when the last step from *EI* to *E* + *P* is much faster than the previous step, we get again the original equation. Mathematically we have then $K_M' \approx K_M$ and $k_{cat} \approx k_3$.

Multi-Substrate Reactions

Multi-substrate reactions follow complex rate equations that describe how the substrates bind and in what sequence. The analysis of these reactions is much simpler if the concentration of substrate A is kept constant and substrate B varied. Under these conditions, the enzyme behaves just like a single-substrate enzyme and a plot of v by [S] gives apparent K_M and V_{max} constants for substrate B. If a set of these measurements is performed at different fixed concentrations of A, these data can be used to work out what the mechanism of the reaction is. For an enzyme that takes two substrates A and B and turns them into two products P and Q, there are two types of mechanism: ternary complex and ping–pong.

Ternary-Complex Mechanisms

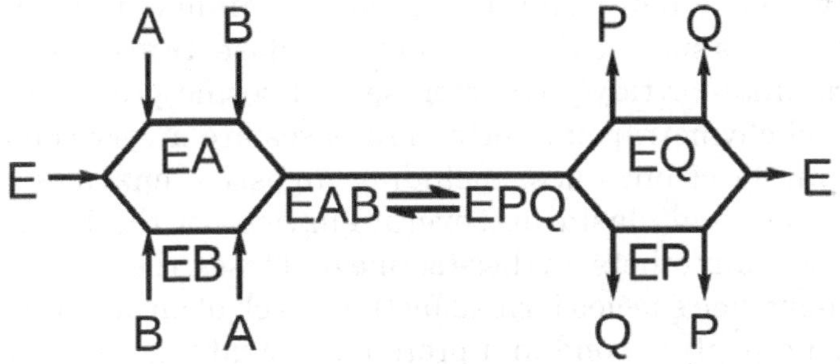

Figure: *Random-order ternary-complex mechanism for an enzyme reaction. The reaction path is shown as a line and enzyme intermediates containing substrates A and B or products P and Q are written below the line.*

In these enzymes, both substrates bind to the enzyme at the same time to produce an EAB ternary complex. The order of binding can either be random (in a random mechanism) or substrates have to bind in a particular sequence (in an ordered mechanism). When a set of v by [S] curves (fixed A, varying B) from an enzyme with a ternary-complex mechanism are plotted in a Lineweaver–Burk plot, the set of lines produced will intersect.

Enzymes with ternary-complex mechanisms include glutathione *S*-transferase, dihydrofolate reductase and DNA polymerase. The following links show short animations of the ternary-complex mechanisms of the enzymes dihydrofolate reductase and DNA polymerase.

Ping–Pong Mechanisms

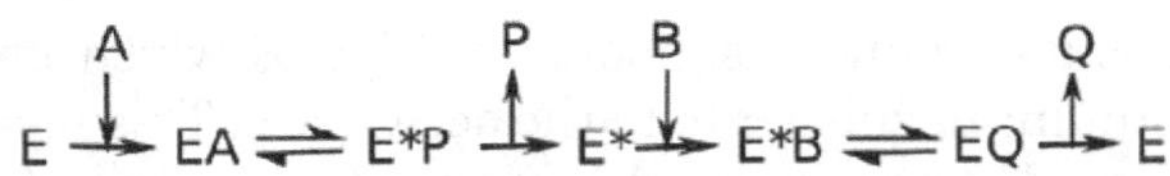

Figure: *Ping–pong mechanism for an enzyme reaction. Intermediates contain substrates A and B or products P and Q.*

As shown on the right, enzymes with a ping-pong mechanism can exist in two states, E and a chemically modified form of the enzyme E*; this modified enzyme is known as an intermediate. In such mechanisms, substrate A binds, changes the enzyme to E* by, for example, transferring a chemical group to the active site, and is then released. Only after the first substrate is released can substrate B bind and react with the modified enzyme, regenerating the unmodified E form. When a set of v by [S] curves (fixed A, varying B) from an enzyme with a ping–pong mechanism are plotted in a Lineweaver–Burk plot, a set of parallel lines will be produced. This is called a secondary plot.

Enzymes with ping–pong mechanisms include some oxidoreductases such as thioredoxin peroxidase, transferases such as acylneuraminate cytidylyltransferase and serine proteases such as trypsin and chymotrypsin. Serine proteases are a very common and diverse family of enzymes, including digestive enzymes (trypsin, chymotrypsin, and elastase), several enzymes of the blood clotting cascade and many others. In these serine proteases, the E* intermediate is an acyl-enzyme species formed by the attack of an active site serine residue on a peptide bond in a protein substrate. A short animation showing the mechanism of chymotrypsin is linked here.

Non-Michaelis–Menten Kinetics

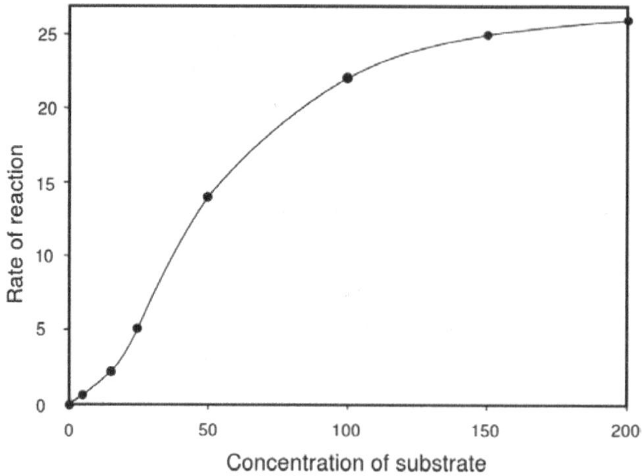

Figure: *Saturation curve for an enzyme reaction showing sigmoid kinetics.*

Some enzymes produce a sigmoid v by [S] plot, which often indicates cooperative binding of substrate to the active site. This means that the binding of one substrate molecule affects the binding of subsequent substrate molecules. This behaviour is most common in multimeric enzymes with several interacting active sites. Here, the mechanism of cooperation is similar to that of hemoglobin, with binding of substrate to one active site altering the affinity of the other active sites for substrate molecules. Positive cooperativity occurs when binding of the first substrate molecule *increases* the affinity of the other active sites for substrate. Negative cooperativity occurs when binding of the first substrate *decreases* the affinity of the enzyme for other substrate molecules.

Allosteric enzymes include mammalian tyrosyl tRNA-synthetase, which shows negative cooperativity, and bacterial aspartate transcarbamoylase and phosphofructokinase, which show positive cooperativity.

Cooperativity is surprisingly common and can help regulate the responses of enzymes to changes in the concentrations of their substrates. Positive cooperativity makes enzymes much more sensitive to [S] and their activities can show large changes over a narrow range of substrate concentration. Conversely, negative cooperativity makes enzymes insensitive to small changes in [S].

The Hill equation (biochemistry) is often used to describe the degree of cooperativity quantitatively in non-Michaelis–Menten kinetics. The derived Hill coefficient *n* measures how much the binding of substrate to one active site affects the binding of substrate to the other active sites. A Hill coefficient of <1 indicates negative cooperativity and a coefficient of >1 indicates positive cooperativity.

Pre-Steady-State Kinetics

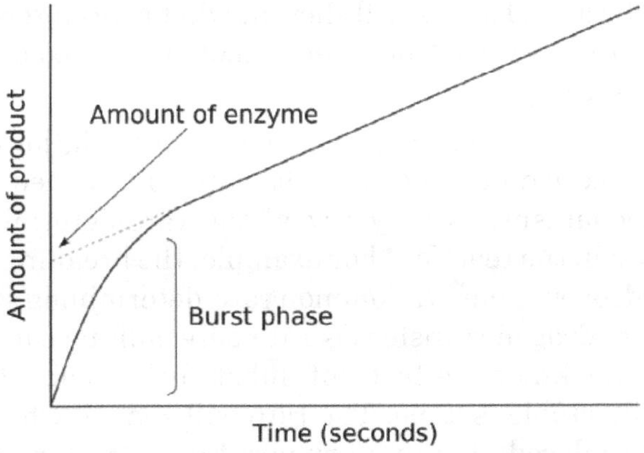

Figure: *Pre-steady state progress curve, showing the burst phase of an enzyme reaction.*

In the first moment after an enzyme is mixed with substrate, no product has been formed and no intermediates exist. The study of the next few milliseconds of the reaction is called Pre-steady-state kinetics also referred to as Burst kinetics. Pre-steady-state kinetics is therefore concerned with the formation and consumption of enzyme–substrate intermediates (such as ES or E*) until their steady-state concentrations are reached.

This approach was first applied to the hydrolysis reaction catalysed by chymotrypsin. Often, the detection of an intermediate is a vital piece of evidence in investigations of what mechanism an enzyme follows. For example, in the ping–pong mechanisms that are shown above, rapid kinetic measurements can follow the release of product P and measure the formation of the modified enzyme intermediate E*. In the case of chymotrypsin, this intermediate is formed by an attack on the substrate by the nucleophilic serine in the active site and the formation of the acyl-enzyme intermediate.

In the figure to the right, the enzyme produces E* rapidly in the first few seconds of the reaction. The rate then slows as steady state is reached. This rapid burst phase of the reaction measures a single turnover of the enzyme. Consequently, the amount of product released in this burst, shown as the intercept on the *y*-axis of the graph, also gives the amount of functional enzyme which is present in the assay.

Chemical Mechanism

An important goal of measuring enzyme kinetics is to determine the chemical mechanism of an enzyme reaction, i.e., the sequence of chemical steps that transform substrate into product. The kinetic approaches discussed above will show at what rates intermediates are formed and inter-converted, but they cannot identify exactly what these intermediates are.

Kinetic measurements taken under various solution conditions or on slightly modified enzymes or substrates often shed light on this chemical mechanism, as they reveal the rate-determining step or intermediates in the reaction. For example, the breaking of a covalent bond to a hydrogen atom is a common rate-determining step. Which of the possible hydrogen transfers is rate determining can be shown by measuring the kinetic effects of substituting each hydrogen by deuterium, its stable isotope. The rate will change when the critical hydrogen is replaced, due to a primary kinetic isotope effect, which occurs because bonds to deuterium are harder to break than bonds to hydrogen. It is also possible to measure similar effects with other isotope substitutions, such as $^{13}C/^{12}C$ and $^{18}O/^{16}O$, but these effects are more subtle.

Isotopes can also be used to reveal the fate of various parts of the substrate molecules in the final products. For example, it is sometimes difficult to discern the origin of an oxygen atom in the final product; since it may have come from water or from part of the substrate. This may be determined by systematically substituting oxygen's stable isotope ^{18}O into the various molecules that participate in the reaction and checking for the isotope in the product. The chemical mechanism can also be elucidated by examining the kinetics and isotope effects under different pH conditions, by altering the metal ions or other bound cofactors, by site-directed mutagenesis of conserved amino acid residues, or by studying the behaviour of the enzyme in the presence of analogues of the substrate(s).

Enzyme Inhibition and Activation

Enzyme inhibitors are molecules that reduce or abolish enzyme activity, while enzyme activators are molecules that increase the catalytic rate of enzymes. These interactions can be either *reversible* (i.e., removal of the inhibitor restores enzyme activity) or *irreversible* (i.e., the inhibitor permanently inactivates the enzyme).

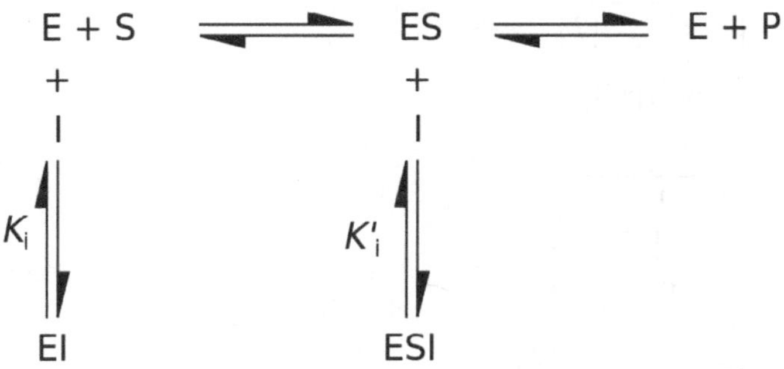

***Figure:** Kinetic scheme for reversible enzyme inhibitors.*

Reversible Inhibitors

Traditionally reversible enzyme inhibitors have been classified as competitive, uncompetitive, or non-competitive, according to their effects on K_m and V_{max}. These different effects result from the inhibitor binding to the enzyme E, to the enzyme–substrate complex ES, or to both, respectively. The division of these classes arises from a problem in their derivation and results in the need to use two different binding constants for one binding event.

The binding of an inhibitor and its effect on the enzymatic activity are two distinctly different things, another problem the traditional equations fail to acknowledge. In noncompetitive inhibition the binding of the inhibitor results in 100% inhibition of the enzyme only, and fails to consider the possibility of anything in between. The common form of the inhibitory term also obscures the relationship between the inhibitor binding to the enzyme and its relationship to any other binding term be it the Michaelis–Menten equation or a dose response curve associated with ligand receptor binding. To demonstrate the relationship the following rearrangement can be made:

$$\frac{V_{\max}}{1+\frac{[I]}{K_i}}$$

$$\frac{V_{\max}}{\frac{[I]+K_i}{K_i}}$$

Adding zero to the bottom ([I]-[I])

$$\frac{V_{\max}}{\frac{[I]+K_i}{[I]+K_i-[I]}}$$

Dividing by $[I]+K_i$

$$\frac{V_{\max}}{\frac{1}{1-\frac{[I]}{[I]+K_i}}}$$

$$V_{\max}-V_{\max}\frac{[I]}{[I]+K_i}$$

This notation demonstrates that similar to the Michaelis–Menten equation, where the rate of reaction depends on the percent of the enzyme population interacting with substrate

fraction of the enzyme population bound by substrate

$$\frac{[S]}{[S]+K_m}$$

fraction of the enzyme population bound by inhibitor

$$\frac{[I]}{[I]+K_i}$$

the effect of the inhibitor is a result of the percent of the enzyme population interacting with inhibitor. The only problem with this equation in its present form is that it assumes absolute inhibition of the enzyme with inhibitor binding, when in fact there can be a wide range of effects anywhere from 100% inhibition of substrate turn over to just >0%. To account for this the equation can be easily modified to allow for different degrees of inhibition by including a delta $V_{\max}$ term.

$$V_{max} - \Delta V_{max} \frac{[I]}{[I]+K_i}$$

or

$$V_{max}1 - (V_{max}1 - V_{max}2)\frac{[I]}{[I]+K_i}$$

This term can then define the residual enzymatic activity present when the inhibitor is interacting with individual enzymes in the population. However the inclusion of this term has the added value of allowing for the possibility of activation if the secondary V_{max} term turns out to be higher than the initial term. To account for the possibly of activation as well the notation can then be rewritten replacing the inhibitor "I" with a modifier term denoted here as "X".

$$V_{max}1 - (V_{max}1 - V_{max}2)\frac{[X]}{[X]+K_x}$$

While this terminology results in a simplified way of dealing with kinetic effects relating to the maximum velocity of the Michaelis–Menten equation, it highlights potential problems with the term used to describe effects relating to the K_m. The K_m relating to the affinity of the enzyme for the substrate should in most cases relate to potential changes in the binding site of the enzyme which would directly result from enzyme inhibitor interactions. As such a term similar to the one proposed above to modulate V_{max} should be appropriate in most situations:

$$K_m1 - (K_m1 - K_m2)\frac{[X]}{[X]+K_x}$$

Irreversible Inhibitors

Enzyme inhibitors can also irreversibly inactivate enzymes, usually by covalently modifying active site residues. These reactions, which may be called suicide substrates, follow exponential decay functions and are usually saturable. Below saturation, they follow first order kinetics with respect to inhibitor.

Mechanisms of Catalysis

The favoured model for the enzyme–substrate interaction is the induced fit model. This model proposes that the initial interaction between enzyme and substrate is relatively weak, but that these weak interactions rapidly induce conformational changes in the enzyme that strengthen binding. These conformational changes also bring catalytic residues in the active site close to the chemical bonds in the substrate that will be altered in the reaction. Conformational changes can be measured using circular dichroism or dual polarisation interferometry.

After binding takes place, one or more mechanisms of catalysis lower the energy of the reaction's transition state by providing an alternative chemical pathway for the reaction. Mechanisms of catalysis include catalysis by bond strain; by proximity and orientation; by active-site proton donors or acceptors; covalent catalysis and quantum tunnelling.

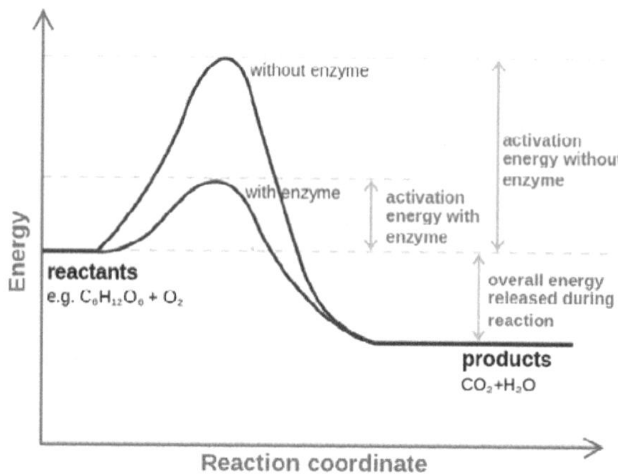

Figure: *The energy variation as a function of reaction coordinate shows the stabilisation of the transition state by an enzyme.*

Enzyme kinetics cannot prove which modes of catalysis are used by an enzyme. However, some kinetic data can suggest possibilities to be examined by other techniques. For example, a ping–pong mechanism with burst-phase pre-steady-state kinetics would suggest covalent catalysis might be important in this enzyme's mechanism. Alternatively, the observation of a strong pH effect on V_{max} but not K_m might indicate that a residue in the active site needs to be in a particular ionisation state for catalysis to occur.

Software

ENZO

ENZO (Enzyme Kinetics) is a graphical interface tool for building kinetic models of enzyme catalyzed reactions. ENZO automatically generates the corresponding differential equations from a stipulated enzyme reaction scheme. These differential equations are processed by a numerical solver and a regression algorithm which fits the coefficients of differential equations to experimentally observed time course curves. ENZO allows rapid evaluation of rival reaction schemes and can be used for routine tests in enzyme kinetics.

Protein Dynamics

Proteins are generally thought to adopt unique structures determined by their amino acid sequences, as outlined by Anfinsen's dogma. However, proteins are not strictly static objects, but rather populate ensembles of (sometimes similar) conformations.

Transitions between these states occur on a variety of length scales (tenths of Å to nm) and time scales (ns to s), and have been linked to functionally relevant phenomena such as allosteric signaling and enzyme catalysis.

The study of protein dynamics is most directly concerned with the transitions between these states, but can also involve the nature and equilibrium populations of the states themselves. These two perspectives—kinetics and thermodynamics, respectively—can be conceptually synthesized in an "energy landscape" paradigm: highly populated states and the kinetics of transitions between them can be described by the depths of energy wells and the heights of energy barriers, respectively.

Local Flexibility: Atoms and Residues

Portions of protein structures often deviate from the equilibrium state. Some such excursions are harmonic, such as stochastic fluctuations of chemical bonds and bond angles. Others are anharmonic, such as sidechains that jump between separate discrete energy minima, or rotamers.

Evidence for local flexibility is often obtained from NMR spectroscopy with methods like Random Coil Index. However, it can also be derived from very high-resolution electron density maps produced by X-ray crystallography, particularly when diffraction data is collected at room temperature instead of the traditional cryogenic temperature (typically near 100 K).

Regional Flexibility: Intra-Domain Multi-Residue Coupling

Many residues are in close spatial proximity in protein structures. This is true for most residues that are contiguous in the primary sequence, but also for many that are distal in sequence yet are brought into contact in the final folded structure. Because of this proximity, these residues's energy landscapes become coupled based on various biophysical phenomena such as hydrogen bonds, ionic bonds, and van der Waals interactions. Transitions between states for such sets of residues therefore become correlated.

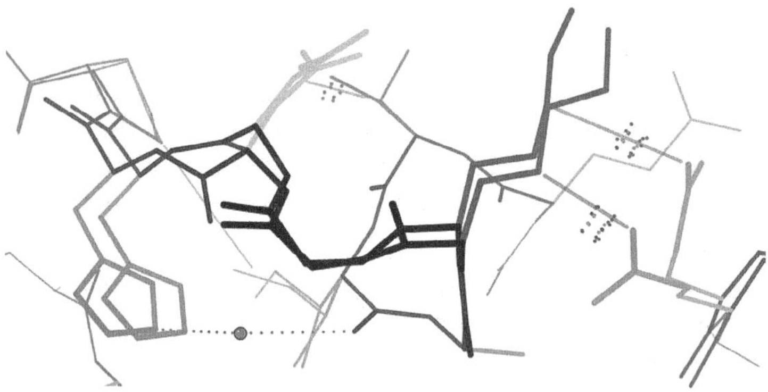

Figure: *A network of alternative conformations in catalase (Protein Data Bank code: 1gwe) with diverse properties. Multiple phenomena define the network: van der Waals interactions (blue dots and line segments) between sidechains, a hydrogen bond (dotted green line) through a partial-occupancy water (brown), coupling through the locally mobile backbone (black), and perhaps electrostatic forces between the Lys (green) and nearby polar residues (blue: Glu, yellow: Asp, purple: Ser). This particular network is distal from the active site and is therefore putatively not critical for function.*

This is perhaps most obvious for surface-exposed loops, which often shift collectively to adopt different conformations in different crystal structures. However, coupled conformational heterogeneity is also sometimes evident in secondary structure.

For example, consecutive residues and residues offset by 4 in the primary sequence often interact in α helices. Also, residues offset by 2 in the primary sequence point their sidechains toward the same face of β sheets and are close enough to interact sterically, as are residues on adjacent strands of the same β sheet.

When these coupled residues form pathways linking functionally important parts of a protein, they may participate in allosteric signaling. For example, when a molecule of oxygen binds to one subunit of the hemoglobin tetramer, that information is allosterically propagated to the other three subunits, thereby enhancing their affinity for oxygen.

In this case, the coupled flexibility in hemoglobin allows for cooperative oxygen binding, which is physiologically useful because it allows rapid oxygen loading in lung tissue and rapid oxygen unloading in oxygen-deprived tissues (e.g. muscle).

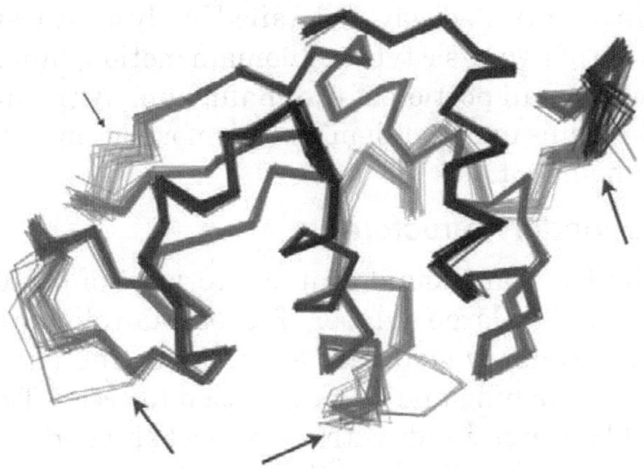

Figure: *An "ensemble" of 44 crystal structures of hen egg white lysozyme from the Protein Data Bank, showing that different crystallization conditions lead to different conformations for various surface-exposed loops and termini (red arrows).*

Global Flexibility: Multiple Domains

The presence of multiple domains in proteins gives rise to a great deal of flexibility and mobility, leading to protein domain dynamics. Domain motions can be inferred by comparing different structures of a protein (as in Database of Molecular Motions), or they can be directly observed using spectra measured by neutron spin echo spectroscopy. They can also be suggested by sampling in extensive molecular dynamics trajectories and principal component analysis. Domain motions are important for:

- catalysis
- regulatory activity
- transport of metabolites
- formation of protein assemblies
- cellular locomotion

One of the largest observed domain motions is the 'swivelling' mechanism in pyruvate phosphate dikinase. The phosphoinositide domain swivels between two states in order to bring a phosphate group from the active site of the nucleotide binding domain to that of the phosphoenolpyruvate/pyruvate domain. The phosphate group is moved over a distance of 45 Å involving a domain motion of about 100 degrees around a single residue. In enzymes, the closure of one domain onto another captures a substrate by an induced fit, allowing the reaction to take place in a controlled way. A detailed analysis by Gerstein led to the classification of two basic types of domain motion; hinge and shear. Only a relatively small portion of the chain, namely the inter-domain linker and side chains undergo significant conformational changes upon domain rearrangement.

Hinges by Secondary Structures

A study by Hayward found that the termini of α-helices and β-sheets form hinges in a large number of cases. Many hinges were found to involve two secondary structure elements acting like hinges of a door, allowing an opening and closing motion to occur. This can arise when two neighbouring strands within a β-sheet situated in one domain, diverge apart as they join the other domain. The two resulting termini then form the bending regions between the two domains. α-helices that preserve their hydrogen bonding network when bent are found to behave as mechanical hinges, storing 'elastic energy' that drives the closure of domains for rapid capture of a substrate.

Helical to Extended Conformation

The interconversion of helical and extended conformations at the site of a domain boundary is not uncommon. In calmodulin, torsion angles change for five residues in the middle of a domain linking α-helix. The helix is split into two, almost perpendicular, smaller helices separated by four residues of an extended strand.

Shear Motions

Shear motions involve a small sliding movement of domain interfaces, controlled by the amino acid side chains within the interface. Proteins displaying shear motions often have a layered architecture: stacking of secondary structures. The interdomain linker has merely the role of keeping the domains in close proximity.

Domain Motion and Functional Dynamics in Enzymes

The analysis of the internal dynamics of structurally different, but functionally similar enzymes has highlighted a common relationship between the positioning of the active site and the two principal protein sub-domains. In fact, for several members of the hydrolase superfamily, the catalytic site is located close to the interface separating the two principal quasi-rigid domains. Such positioning appears instrumental for maintaining the precise geometry of the active site, while allowing for an appreciable functionally oriented modulation of the flanking regions resulting from the relative motion of the two sub-domains.

Implications for Macromolecular Evolution

Evidence suggests that protein dynamics are important for function, e.g. enzyme catalysis in DHFR, yet they are also posited to facilitate the acquisition of new functions by molecular evolution. This argument suggests that proteins have evolved to have stable, mostly unique folded structures, but the unavoidable residual flexibility leads to some degree of functional promiscuity, which can be amplified/ harnessed/diverted by subsequent mutations.

However, there is growing awareness that intrinsically unstructured proteins are quite prevalent in eukaryotic genomes, casting further doubt on the simplest interpretation of Anfinsen's dogma: "sequence determines structure (singular)". In effect, the new paradigm is characterized by the addition of two caveats: "sequence and cellular environment determine structural ensemble".

Chapter 2

Chemical Binding

A chemical bond is an attraction between atoms that allows the formation of chemical substances that contain two or more atoms. The bond is caused by the electrostatic force of attraction between opposite charges, either between electrons and nuclei, or as the result of a dipole attraction.

The strength of chemical bonds varies considerably; there are "strong bonds" such as covalent or ionic bonds and "weak bonds" such as dipole–dipole interactions, the London dispersion force and hydrogen bonding.

Since opposite charges attract via a simple electromagnetic force, the negatively charged electrons that are orbiting the nucleus and the positively charged protons in the nucleus attract each other. An electron positioned between two nuclei will be attracted to both of them, and the nuclei will be attracted toward electrons in this position.

This attraction constitutes the chemical bond. Due to the matter wave nature of electrons and their smaller mass, they must occupy a much larger amount of volume compared with the nuclei, and this volume occupied by the electrons keeps the atomic nuclei relatively far apart, as compared with the size of the nuclei themselves. This phenomenon limits the distance between nuclei and atoms in a bond.

In general, strong chemical bonding is associated with the sharing or transfer of electrons between the participating atoms.

The atoms in molecules, crystals, metals and diatomic gases—indeed most of the physical environment around us— are held together by chemical bonds, which dictate the structure and the bulk properties of matter.

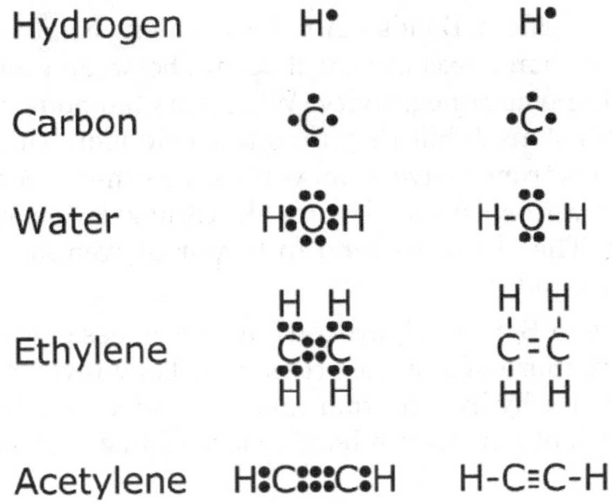

Figure: *Examples of Lewis dot-style representations of chemical bonds between carbon (C), hydrogen (H), and oxygen (O). Lewis dot diagrams were an early attempt to describe chemical bonding and are still widely used today.*

Understanding Biology

Chemical bonds play an important role in the world of biology. Chemical bonds hold your computer together, the cells in your body together, and connect almost everything around us on the atomic level. A chemical bond is an attraction between atoms due to sharing of electrons between atoms or a complete transfer of electrons from one atom to another. There are three types of chemical bonds. They are ionic, covalent, and polar covalent. Separate from chemical bonds are also hydrogen bonds, which deal with hydrogen atoms.

A Review of the Types of Chemical Bonds

- Ionic Bonds - An ionic bond is a complete transfer of electrons from one atom to another. This generally happens between atoms that have opposite electronegativity. This means one has very few atoms in their outer shell, while the other has many. A common example of an ionic bond is that of salt, with Na and Cl. Sodium has one electron in its outer shell, in which it transfers to chloride to make an ionic bond.
- Covalent Bonds - Covalent bonds involve a complete sharing of electrons between two atoms. It occurs most commonly between atoms that have outer shells that are only partially filled. If the two atoms have similar electronegativity, then the electrons can be shared between them. Carbon forms covalent bonds.

- Polar Covalent Bonds - A polar covalent bond is much like a covalent bond, except that it occurs between atoms that have differing electronegativity. When this happens, the electrons are still shared, but they tend to spend more time around the more electronegative atom versus the other. Such an example is with water. Oxygen is very electronegative, while hydrogen is not. The electrons tend to favour oxygen and spend more time around it.
- Hydrogen Bonds - Hydrogen bonds are less of chemical bonds and are more of a static attraction. They involve the reaction between a hydrogen atom and an electronegative atom. The strength of a hydrogen bond is less than a tenth of a covalent bond.

Ionic Bonds in Biology

Ionic bonds are usually found in dry forms such as salts and are found in compounds throughout the human body. Ionic compounds are generally water soluble.

- Protein Shapes - Ionic bonds play an important role in shaping tertiary and quaternary proteins. As a protein undergoes its folding process, certain atoms on the protein will be attracted to one another and will from ionic bonds that hold the general shape of a protein together.
- Chromosome Shapes - Much like protein shapes, ionic bonds help determine the shapes of chromosomes depending on what atoms bond to each other.
- Muscle Contraction - There are many different ionic compounds (salts) in cells. Ions are used to maintain cell potentials and are important in cell signalling and muscle contraction. As ions move in and out of cells, they can build up positive and negative charges on different sides of those cells. Once these charges build up, they can trigger a reaction such as muscle or nerve contraction.

Ionic bonds are also involved in determining cell shapes, catalytic reactions, and neuron functions.

Covalent & Polar Covalent Bonds in Biology

Covalent bonds can be found as gasses, liquids, or solids and are all around us. They are not soluble in water. The atoms of materials with covalent bonds are bound tightly to each other in stable molecules, yet, they are generally not very attracted to other molecules in or around the material. Carbon-Carbon bonds (C-C) are covalent bonds and form the basis of most biomolecules.

Sugars: Monosaccharides such as glucose, fructose and galactose are held together by a type of covalent bond known as a glycosidic bond. When they are connected together, they can form larger carbohydrates. Two monosaccharides linked together form a disaccharide. Such an example is sucrose, which is composed of two monosaccharides, glucose and fructose.

Polar covalent bonds link together amino acids in chains, creating peptides. They can also be found holding together amines.

Hydrogen Bonds In Biology

Hydrogen bonds are considerably weak when compared to others, but these weak chemical interactions are very important in the world of biology, dues to the fact that the molecules they form can be easily taken apart and put back together again. Some of these hydrogen bonds are significant in the survival of organisms.

- Water - The hydrogen on a water molecule has a positive charge, while oxygen has a negative one. The opposite ends of water molecules attract each other simply with negative and positive attraction. This is why water droplets are attracted to each other when they are in a small distance of each other. Water can form this type of bond with other polar molecules or ions such as sodium ions or hydrogen ions.
- Ice - When water freezes and turns to ice, hydrogen bonds space water molecules further apart than they would normally be in the liquid form. Because of the extra space now created, water is less dense when it freezes, and it enables ice to float on top of the liquid form of water.
- Water Striders - The water strider is an insect that hunts for its food on the surface of still water, such as a pond or swamp. It has widely spaced feet that depress the surface of the water when they are spread out, creating a weight light enough to be held up by the surface tension created from the hydrogen bonding in the water. This essentially helps it to walk on water.
- DNA - As you recall, DNA is formed in the shape of a double helix. The double strands of DNA are held together by hydrogen bonds. Each single strand has a backbone made of sugar and phosphate, as well as either a purine (adenine or guanine) or pyrimidine (cytosine or thymine). Each purine is connected to

a pyrimidine through a hydrogen bond, giving the double DNA strand strength, and flexibility. This bond holds the two sides of DNA together, each bond contributing to the overall strength of DNA. When DNA is replicated, special enzymes known as DNA helicase "unzip" DNA and these bonds are broken so the two strands can be individually replicated.

After studying chemical bonds and their applications in biology, it becomes clear to us that the world around us is connected at so many levels, and it makes sense that all sciences are linked in some way.

Overview of Main Types of Chemical Bonds

A chemical bond is an attraction between atoms. This attraction may be seen as the result of different behaviours of the outermost electrons of atoms. Although all of these behaviours merge into each other seamlessly in various bonding situations so that there is no clear line to be drawn between them, the behaviours of atoms become so qualitatively different as the character of the bond changes quantitatively, that it remains useful and customary to differentiate between the bonds that cause these different properties of condensed matter.

In the simplest view of a so-called 'covalent' bond, one or more electrons (often a pair of electrons) are drawn into the space between the two atomic nuclei. Here the negatively charged electrons are attracted to the positive charges of *both* nuclei, instead of just their own. This overcomes the repulsion between the two positively charged nuclei of the two atoms, and so this overwhelming attraction holds the two nuclei in a fixed configuration of equilibrium, even though they will still vibrate at equilibrium position.

Thus, covalent bonding involves sharing of electrons in which the positively charged nuclei of two or more atoms simultaneously attract the negatively charged electrons that are being shared between them. These bonds exist between two particular identifiable atoms, and have a direction in space, allowing them to be shown as single connecting lines between atoms in drawings, or modelled as sticks between spheres in models. In a polar covalent bond, one or more electrons are unequally shared between two nuclei. Covalent bonds often result in the formation of small collections of better-connected atoms called molecules, which in solids and liquids are bound to other molecules by forces that are often much weaker than the covalent bonds that hold the molecules internally together. Such weak intermolecular bonds give organic molecular substances, such as waxes and oils, their soft bulk character, and their low melting points (in liquids, molecules must cease most structured or oriented contact with each other). When covalent bonds link long chains of atoms in large molecules, however (as in polymers such as nylon), or when covalent bonds extend in networks through solids that are not composed of discrete molecules (such as diamond or quartz or the silicate minerals in many types of rock) then the structures that result may be both strong and tough, at least in the direction oriented correctly with networks of covalent bonds. Also, the melting points of such covalent polymers and networks increase greatly.

In a simplified view of an *ionic* bond, the bonding electron is not shared at all, but transferred. In this type of bond, the outer atomic orbital of one atom has a vacancy which allows addition of one or more electrons. These newly added electrons potentially occupy a lower energy-state (effectively closer to more nuclear charge) than they experience in a different atom. Thus, one nucleus offers a more tightly bound position to an electron than does another nucleus, with the result that one atom may transfer an electron to the other. This transfer causes one atom to assume a net positive charge, and the other to assume a net negative charge.

The *bond* then results from electrostatic attraction between atoms, and the atoms become positive or negatively charged ions. Ionic bonds may be seen as extreme examples of polarization in covalent bonds. Often, such bonds have no particular orientation in space, since they result from equal electrostatic attraction of each ion to all ions around them. Ionic bonds are strong (and thus ionic substances require high temperatures to melt) but also brittle, since the forces between ions are short-range, and do not easily bridge cracks and fractures. This type of bond gives rise to the physical characteristics of crystals of classic mineral salts, such as table salt.

A less often mentioned type of bonding is the *metallic* bond. In this type of bonding, each atom in a metal donates one or more electrons to a "sea" of electrons that reside between many metal atoms. In this sea, each electron is free (by virtue of its wave nature) to be associated with a great many atoms at once. The bond results because the metal atoms become somewhat positively charged due to loss of their electrons, while the electrons remain attracted to many atoms, without being part of any given atom. Metallic bonding may be seen as an extreme example of delocalization of electrons over a large system of covalent bonds, in which every atom participates. This type of bonding is often very strong (resulting in the tensile strength of metals). However, metallic bonds are more collective in nature than other types, and so they allow metal crystals to more easily deform, because they are composed of atoms attracted to each other, but not in any particularly-oriented ways. This results in the malleability of metals. The sea of electrons in metallic bonds causes the characteristically good electrical and thermal conductivity of metals, and also their "shiny" reflection of most frequencies of white light.

All bonds can be explained by quantum theory, but, in practice, simplification rules allow chemists to predict the strength, directionality, and polarity of bonds. The octet rule and VSEPR theory are two examples. More sophisticated theories are valence bond theory which includes orbital hybridization and resonance, and the linear combination of atomic orbitals molecular orbital method which includes ligand field theory. Electrostatics are used to describe bond polarities and the effects they have on chemical substances.

History

Early speculations into the nature of the chemical bond, from as early as the 12th century, supposed that certain types of chemical species were joined by a type of chemical affinity. In 1704, Isaac Newton famously outlined his atomic bonding theory, in "Query 31" of his Opticks, whereby atoms attach to each other by some "force". Specifically, after acknowledging the various popular theories in vogue at the time, of how atoms were reasoned to attach to each other, i.e. "hooked atoms", "glued together by rest", or "stuck together by conspiring motions", Newton states that he would rather infer from their cohesion, that "particles attract one another by some force, which in immediate contact is exceedingly strong, at small distances performs the chemical operations, and reaches not far from the particles with any sensible effect."

In 1819, on the heels of the invention of the voltaic pile, Jöns Jakob Berzelius developed a theory of chemical combination stressing the electronegative and electropositive character of the combining atoms. By the mid 19th century, Edward Frankland, F.A. Kekulé, A.S. Couper, Alexander Butlerov, and Hermann Kolbe, building on the theory of radicals, developed the theory of valency, originally called "combining power", in which compounds were joined owing to an attraction of positive and negative poles. In 1916, chemist Gilbert N. Lewis developed the concept of the electron-pair bond, in which two atoms may share one to six electrons, thus forming the single electron bond, a single bond, a double bond, or a triple bond; in Lewis's own words, "An electron may form a part of the shell of two different atoms and cannot be said to belong to either one exclusively."

That same year, Walther Kossel put forward a theory similar to Lewis' only his model assumed complete transfers of electrons between atoms, and was thus a model of ionic bonds. Both Lewis and Kossel structured their bonding models on that of Abegg's rule (1904).

In 1927, the first mathematically complete quantum description of a simple chemical bond, i.e. that produced by one electron in the hydrogen molecular ion, H_2^+, was derived by the Danish physicist Oyvind Burrau. This work showed that the quantum approach to chemical bonds could be fundamentally and quantitatively correct, but the mathematical methods used could not be extended to molecules containing more than one electron. A more practical, albeit less quantitative, approach was put forward in the same year by Walter Heitler and Fritz London.

The Heitler-London method forms the basis of what is now called valence bond theory. In 1929, the linear combination of atomic orbitals molecular orbital method (LCAO) approximation was introduced by Sir John Lennard-Jones, who also suggested methods to derive electronic structures of molecules of F_2 (fluorine) and O_2 (oxygen) molecules, from basic quantum principles. This molecular orbital theory represented a covalent bond as an orbital formed by combining the quantum mechanical Schrödinger atomic orbitals which had been hypothesized for electrons in single atoms. The equations for bonding electrons in multi-electron atoms could not be solved to mathematical perfection (i.e., *analytically*), but approximations for them still gave many good qualitative predictions and results. Most quantitative calculations in modern quantum chemistry use either valence bond or molecular orbital theory as a starting point, although a third approach, density functional theory, has become increasingly popular in recent years.

In 1933, H. H. James and A. S. Coolidge carried out a calculation on the dihydrogen molecule that, unlike all previous calculation which used functions only of the distance of the electron from the atomic nucleus, used functions which also explicitly added the distance between the two electrons. With up to 13 adjustable parameters they obtained a result very close to the experimental result for the dissociation energy. Later extensions have used up to 54 parameters and give excellent agreement with experiment. This calculation convinced the scientific community that quantum theory could give agreement with experiment. However this approach has none of the physical pictures of the valence bond and molecular orbital theories and is difficult to extend to larger molecules.

Valence Bond Theory

In 1927, valence bond theory was formulated and it argues that a chemical bond forms when two valence electrons, in their respective atomic orbitals, work or function to hold two nuclei together, by virtue of effects of lowering system energies.

Building on this theory, the chemist Linus Pauling published in 1931 what some consider one of the most important papers in the history of chemistry: "On the Nature of the Chemical Bond". In this paper, elaborating on the works of Lewis, and the valence bond theory (VB) of Heitler and London, and his own earlier works, Pauling presented six rules for the shared electron bond, the first three of which were already generally known:

1. The electron-pair bond forms through the interaction of an unpaired electron on each of two atoms.
2. The spins of the electrons have to be opposed.
3. Once paired, the two electrons cannot take part in additional bonds.

His last three rules were new:

4. The electron-exchange terms for the bond involves only one wave function from each atom.
5. The available electrons in the lowest energy level form the strongest bonds.
6. Of two orbitals in an atom, the one that can overlap the most with an orbital from another atom will form the strongest bond, and this bond will tend to lie in the direction of the concentrated orbital.

Building on this article, Pauling's 1939 textbook: *On the Nature of the Chemical Bond* would become what some have called the "Bible" of modern chemistry. This book helped experimental chemists to understand the impact of quantum theory on chemistry.

However, the later edition in 1959 failed to adequately address the problems that appeared to be better understood by molecular orbital theory. The impact of valence theory declined during the 1960s and 1970s as molecular orbital theory grew in usefulness as it was implemented in large digital computer programs. Since the 1980s, the more difficult problems of implementing valence bond theory into computer programs have been solved largely, and valence bond theory has seen a resurgence.

Comparison of Valence Bond and Molecular Orbital Theory

In some respects valence bond theory is superior to molecular orbital theory. When applied to the simplest two-electron molecule, H_2, valence bond theory, even at the simplest Heitler-London approach, gives a much closer approximation to the bond energy, and it provides a much more accurate representation of the behaviour of the electrons as chemical bonds are formed and broken. In contrast simple molecular orbital theory predicts that the hydrogen molecule dissociates into a linear superposition of hydrogen atoms and positive and negative hydrogen ions, a completely unphysical result. This explains in part why the curve of total energy against interatomic distance for the valence bond method lies below the curve for the molecular orbital method at all distances and most particularly so for large distances. This situation arises for all homonuclear diatomic molecules and is particularly a problem for F_2, where the minimum energy of the curve with molecular orbital theory is still higher in energy than the energy of two F atoms.

The concepts of hybridization are so versatile, and the variability in bonding in most organic compounds is so modest, that valence bond theory remains an integral part of the vocabulary of organic chemistry. However, the work of Friedrich Hund, Robert Mulliken, and Gerhard Herzberg showed that molecular orbital theory provided a more appropriate description of the spectroscopic, ionization and magnetic properties of molecules. The deficiencies of valence bond theory became apparent when hypervalent molecules (e.g. PF_5) were explained without the use of d orbitals that were crucial to the bonding hybridisation scheme proposed for such molecules by Pauling. Metal complexes and electron deficient compounds (e.g. diborane) also appeared to be well described by molecular orbital theory, although valence bond descriptions have been made.

In the 1930s the two methods strongly competed until it was realised that they are both approximations to a better theory. If we take the simple valence bond structure and mix in all possible covalent and ionic structures arising from a particular set of atomic orbitals, we reach what is called the full configuration interaction wave function. If we take the simple molecular orbital description of the ground state and combine that function with the functions describing all possible excited states using unoccupied orbitals arising from the same set of atomic orbitals, we also reach the full configuration interaction wavefunction. It can be then seen that the simple molecular orbital approach gives too much weight to the ionic structures, while the simple valence bond approach gives too little. This can also be described as saying that the molecular orbital approach is too *delocalised*, while the valence bond approach is too *localised.*

The two approaches are now regarded as complementary, each providing its own insights into the problem of chemical bonding. Modern calculations in quantum chemistry usually start from (but ultimately go far beyond) a molecular orbital rather than a valence bond approach, not because of any intrinsic superiority in the former but rather because the MO approach is more readily adapted to numerical computations. However better valence bond programs are now available.

Bonds in Chemical Formulas

The fact that atoms and molecules are three-dimensional makes it difficult to use a single technique for indicating orbitals and bonds. In molecular formulas the chemical bonds (binding orbitals) between atoms are indicated by various methods according to the type of discussion. Sometimes, they are completely neglected. For example, in organic chemistry chemists are sometimes concerned only with the functional groups of the molecule.

Thus, the molecular formula of ethanol may be written in a paper in conformational, three-dimensional, full two-dimensional (indicating every bond with no three-dimensional directions), compressed two-dimensional (CH_3–CH_2–OH), separating the functional group from another part of the molecule (C_2H_5OH), or by its atomic constituents (C_2H_6O), according to what is discussed. Sometimes, even the non-bonding valence shell electrons (with the two-dimensional approximate directions) are marked, i.e. for elemental carbon 'C'. Some chemists may also mark the respective orbitals, i.e. the hypothetical ethene"4 anion (\/C=C/\ "4) indicating the possibility of bond formation.

Strong Chemical Bonds

Typical bond lengths in pmand bond energies in kJ/mol. Bond lengths can be converted to Åby division by 100 (1 Å = 100 pm). Data taken from.

Bond	*Length(pm)*	*Energy(kJ/mol)*
	H — Hydrogen	
H–H	74	436
H–0	96	366
H–F	92	568
H–Cl	127	432

Contd...

Bond	***Length(pm)***	***Energy(kJ/mol)***
	C — Carbon	
C–H	109	413
C–C	154	348
C–C=	151	
=C–Ca”	147	
=C–C=	148	
C=C	134	614
Ca”C	120	839
C–N	147	308
C–O	143	360
C–F	134	488
C–Cl	177	330
	N — Nitrogen	
N–H	101	391
N–N	145	170
Na”N	110	945
O — Oxygen		
O–O	148	145
O=O	121	498
	F, Cl, Br, I — Halogens	
F–F	142	158
Cl–Cl	199	243
Br–H	141	366
Br–Br	228	193
I–H	161	298
I–I	267	151

Strong chemical bonds are the *intramolecular* forces which hold atoms together in molecules. A strong chemical bond is formed from the transfer or sharing of electrons between atomic centres and relies on the electrostatic attraction between the protons in nuclei and the electrons in the orbitals. Although these bonds typically involve the transfer of integer numbers of electrons (this is the bond order, which represents one transferred electron or two shared electrons), some systems can have intermediate numbers of bonds. An example of this is the organic molecule benzene, where the bond order is 1.5 for each carbon atom, meaning that it has 1.5 bonds (shares three electrons) with each one of its two neighbours.

The types of strong bond differ due to the difference in electronegativity of the constituent elements. A large difference in electronegativity leads to more polar (ionic) character in the bond.

Ionic Bond

Ionic bonding is a type of electrostatic interaction between atoms which have a large electronegativity difference. There is no precise value that distinguishes ionic from covalent bonding, but a difference of electronegativity of over 1.7 is likely to be ionic, and a difference of less than 1.7 is likely to be covalent. Ionic bonding leads to separate positive and negative ions. Ionic charges are commonly between -3e to +3e. Ionic bonding commonly occurs in metal salts such as sodium chloride (table salt).

A typical feature of ionic bonds is that the species form into ionic crystals, in which no ion is specifically paired with any single other ion, in a specific directional bond. Rather, each species of ion is surrounded by ions of the opposite charge, and the spacing between it and each of the oppositely charged ions near it, is the same for all surrounding atoms of the same type. It is thus no longer possible to associate an ion with any specific other single ionized atom near it. This is a situation unlike that in covalent crystals, where covalent bonds between specific atoms are still discernible from the shorter distances between them, as measured via such techniques as X-ray diffraction.

Ionic crystals may contain a mixture of covalent and ionic species, as for example salts of complex acids, such as sodium cyanide, NaCN. Many minerals are of this type. X-ray diffraction shows that in NaCN, for example, the bonds between sodium cations (Na^+) and the cyanide anions (CN^-) are *ionic*, with no sodium ion associated with any particular cyanide. However, the bonds between C and N atoms in cyanide are of the *covalent* type, making each of the carbon and nitrogen associated with *just one* of its opposite type, to which it is physically much closer than it is to other carbons or nitrogens in a sodium cyanide crystal.

When such crystals are melted into liquids, the ionic bonds are broken first because they are non-directional and allow the charged species to move freely. Similarly, when such salts dissolve into water, the ionic bonds are typically broken by the interaction with water, but the covalent bonds continue to hold. For example, in solution, the cyanide ions, still bound together as single CN^- ions, move independently through the solution, as do sodium ions, as Na^+. In water, charged ions move apart because each of them are more strongly attracted to a number of water molecules, than to each other.

The attraction between ions and water molecules in such solutions is due to a type of weak dipole-dipole type chemical bond. In melted ionic compounds, the ions continue to be attracted to each other, but not in any ordered or crystalline way.

Covalent Bond

Covalent bonding is a common type of bonding, in which the electronegativity difference between the bonded atoms is small or nonexistent. Bonds within most organic compounds are described as covalent.

A polar covalent bond is a covalent bond with a significant ionic character. This means that the electrons are closer to one of the atoms than the other, creating an imbalance of charge. They occur as a bond between two atoms with moderately different electronegativities, and give rise to dipole-dipole interactions. The electronegativity of these bonds is 0.3 to 1.7 .

A coordinate covalent bond is one where both bonding electrons are from one of the atoms involved in the bond. These bonds give rise to Lewis acids and bases. The electrons are shared roughly equally between the atoms in contrast to ionic bonding. Such bonding occurs in molecules such as the ammonium ion (NH_4^+) and are shown by an arrow pointing to the Lewis acid. Also known as non-polar covalent bond, the electronegativity of these bonds range from 0 to 0.3.

Molecules which are formed primarily from non-polar covalent bonds are often immiscible in water or other polar solvents, but much more soluble in non-polar solvents such as hexane.

One- and Three-Electron Bonds

Bonds with one or three electrons can be found in radical species, which have an odd number of electrons. The simplest example of a 1-electron bond is found in the dihydrogen cation, H_2^+. One-electron bonds often have about half the bond energy of a 2-electron bond, and are therefore called "half bonds". However, there are exceptions: in the case of dilithium, the bond is actually stronger for the 1-electron Li_2^+ than for the 2-electron Li_2. This exception can be explained in terms of hybridization and inner-shell effects.

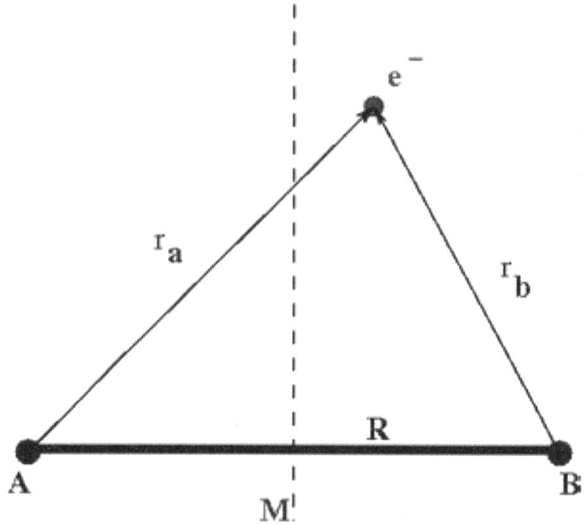

Figure: *One-electron bonding in the dihydrogen cation.*

The simplest example of three-electron bonding can be found in the helium dimer cation, He_2^+. It is considered a "half bond" because it consists of only one shared electron (rather than two) in addition to one lone electron on each atom; in molecular orbital terms, the third electron is in an anti-bonding orbital which cancels out half of the bond formed by the other two electrons. Another example of a molecule containing a 3-electron bond, in addition to two 2-electron bonds, is nitric oxide, NO. The oxygen molecule, O_2 can also be regarded as having two 3-electron bonds and one 2-electron bond, which accounts for its paramagnetism and its formal bond order of 2. Chlorine dioxide and its heavier analogues bromine dioxide and iodine dioxide also contain three-electron bonds.

Molecules with odd-electron bonds are usually highly reactive. These types of bond are only stable between atoms with similar electronegativities.

Bent Bonds

Bent bonds are bonds in strained or otherwise sterically hindered molecules whose binding orbitals are forced into a banana-like form. Bent bonds are often more susceptible to reactions than ordinary bonds.

Resonant Bonding

Hypervalent Bonding

In hypervalent molecules, there exists bonds which have significant non-bonding ionic quality to them. This manifests as non-bonding orbital levels in molecular orbital theory, while in valence bond theory it is analyzed as a form of resonant bonding.

Electron-Deficient Bonding

In three-centre two-electron bonds (-3c–2e") three atoms share two electrons in bonding. This type of bonding occurs in electron deficient compounds like diborane. Each such bond (2 per molecule in diborane) contains a pair of electrons which connect the boron atoms to each other in a banana shape, with a proton (nucleus of a hydrogen atom) in the middle of the bond, sharing electrons with both boron atoms. In certain cluster compounds, so-called four-centre two-electron bonds also have been postulated.

In certain conjugated π (pi) systems, such as benzene and other aromatic compounds, and in conjugated network solids such as graphite, the electrons in the conjugated system of π-bonds are spread over as many nuclear centres as exist in the molecule, or the network.

Aromatic Bonding

In organic chemistry, certain configurations of electrons and orbitals infer extra stability to a molecule. This occurs when π orbitals overlap and combine with others on different atomic centres, forming a long range bond. For a molecule to be aromatic, it must obey Hückel's rule, where the number of π electrons fit the formula $4n + 2$, where n is an integer. The bonds involved in the aromaticity are all planar.

In benzene, the prototypical aromatic compound, 18 ($n = 4$) bonding electrons bind 6 carbon atoms together to form a planar ring structure. The bond "order" (average number of bonds) between the different carbon atoms may be said to be $(18/6)/2=1.5$, but in this case the bonds are all identical from the chemical point of view. They may sometimes be written as single bonds alternating with double bonds, but the view of all ring bonds as being equivalently about 1.5 bonds in strength, is much closer to truth.

In the case of heterocyclic aromatics and substituted benzenes, the electronegativity differences between different parts of the ring may dominate the chemical behaviour of aromatic ring bonds, which otherwise are equivalent.

Metallic Bond

In a metallic bond, bonding electrons are delocalized over a lattice of atoms. By contrast, in ionic compounds, the locations of the binding electrons and their charges are static. The freely-moving or delocalization of bonding electrons leads to classical metallic properties such as luster (surface light reflectivity), electrical and thermal conductivity, ductility, and high tensile strength.

Intermolecular Bonding

There are four basic types of bonds that can be formed between two or more (otherwise non-associated) molecules, ions or atoms. Intermolecular forces cause molecules to be attracted or repulsed by each other. Often, these define some of the physical characteristics (such as the melting point) of a substance.

- A large difference in electronegativity between two bonded atoms will cause a permanent charge separation, or dipole, in a molecule or ion. Two or more molecules or ions with permanent dipoles can interact within dipole-dipole interactions. The bonding electrons in a molecule or ion will, on average, be closer to the more electronegative atom more frequently than the less electronegative one, giving rise to partial charges on each atom, and causing electrostatic forces between molecules or ions.
- A hydrogen bond is effectively a strong example of an interaction between two permanent dipoles. The large difference in electronegativities between hydrogen and any of fluorine, nitrogen and oxygen, coupled with their lone pairs of electrons cause strong electrostatic forces between molecules. Hydrogen bonds are responsible for the high boiling points of water and ammonia with respect to their heavier analogues.
- The London dispersion force arises due to instantaneous dipoles in neighbouring atoms. As the negative charge of the electron is not uniform around the whole atom, there is always a charge imbalance. This small charge will induce a corresponding dipole in a nearby molecule; causing an attraction between the two. The electron then moves to another part of the electron cloud and the attraction is broken.
- A cation–pi interaction occurs between a pi bond and a cation.

Summary: Electrons in Chemical Bonds

In the (unrealistic) limit of "pure" ionic bonding, electrons are perfectly localized on one of the two atoms in the bond. Such bonds can be understood by classical physics. The forces between the atoms are characterized by isotropic continuum electrostatic potentials. Their magnitude is in simple proportion to the charge difference.

Covalent bonds are better understood by valence bond theory or molecular orbital theory. The properties of the atoms involved can be understood using concepts such as oxidation number. The electron density within a bond is not assigned to individual atoms, but is instead delocalized between atoms. In valence bond theory, the two electrons on the two atoms are coupled together with the bond strength depending on the overlap between them. In molecular orbital theory, the linear combination of atomic orbitals (LCAO) helps describe the delocalized molecular orbital structures and energies based on the atomic orbitals of the atoms they came from. Unlike pure ionic bonds, covalent bonds may have directed anisotropic properties. These may have their own names, such as sigma bond and pi bond.

In the general case, atoms form bonds that are intermediates between ionic and covalent, depending on the relative electronegativity of the atoms involved. This type of bond is sometimes called polar covalent.

Energies, Forces and Bonds

In biology, energy is an attribute of all biological systems from the biosphere to the smallest living organism. Within an organism it is responsible for growth and development of a biological cell or an organelle of a biological organism. Energy is thus often said to be stored by cells in the structures of molecules of substances such as carbohydrates (including sugars), lipids, and proteins, which release energy when reacted with oxygen in respiration. In human terms, the human equivalent (H-e) (Human energy conversion) indicates, for a given amount of energy expenditure, the relative quantity of energy needed for human metabolism, assuming an average human energy expenditure of 12,500kJ per day and a basal metabolic rate of 80 watts.

For example, if our bodies run (on average) at 80 watts, then a light bulb running at 100 watts is running at 1.25 human equivalents ($100 \div 80$) i.e. 1.25 H-e. For a difficult task of only a few seconds' duration, a person can put out thousands of watts, many times the 746 watts in one official horsepower. For tasks lasting a few minutes, a fit human can generate perhaps 1,000 watts. For an activity that must be sustained for an hour, output drops to around 300; for an activity kept up all day, 150 watts is about the maximum. The human equivalent assists understanding of energy flows in physical and biological systems by expressing energy units in human terms: it provides a "feel" for the use of a given amount of energy

Sunlight is also captured by plants as *chemical potential energy* in photosynthesis, when carbon dioxide and water (two low-energy compounds) are converted into the high-energy compounds carbohydrates, lipids, and proteins. Plants also release oxygen during photosynthesis, which is utilized by living organisms as an electron acceptor, to release the energy of carbohydrates, lipids, and proteins. Release of the energy stored during photosynthesis as heat or light may be triggered suddenly by a spark, in a forest fire, or it may be made available more slowly for animal or human metabolism, when these molecules are ingested, and catabolism is triggered by enzyme action.

Any living organism relies on an external source of energy—radiation from the Sun in the case of green plants; chemical energy in some form in the case of animals—to be able to grow and reproduce. The daily 1500–2000 Calories (6–8 MJ) recommended for a human adult are taken as a combination of oxygen and food molecules, the latter mostly carbohydrates and fats, of which glucose ($C_6H_{12}O_6$) and stearin ($C_{57}H_{110}O_6$) are convenient examples. The food molecules are oxidised to carbon dioxide and water in the mitochondria

$$C_6H_{12}O_6 + 6O_2 \rightarrow 6CO_2 + 6H_2O$$

$$C_{57}H_{110}O_6 + 81.5O_2 \rightarrow 57CO_2 + 55H_2O$$

and some of the energy is used to convert ADP into ATP

$$ADP + HPO_4^{2-} \rightarrow ATP + H_2O$$

The rest of the chemical energy in the carbohydrate or fat is converted into heat: the ATP is used as a sort of "energy currency", and some of the chemical energy it contains when split and reacted with water, is used for other metabolism (at each stage of a metabolic pathway, some chemical energy is converted into heat). Only a tiny fraction of the original chemical energy is used for work:

gain in kinetic energy of a sprinter during a 100 m race: 4 kJ

gain in gravitational potential energy of a 150 kg weight lifted through 2 metres: 3kJ

Daily food intake of a normal adult: 6–8 MJ

It would appear that living organisms are remarkably inefficient (in the physical sense) in their use of the energy they receive (chemical energy or radiation), and it is true that most real machines manage higher efficiencies. In growing organisms the energy that is converted to heat serves a vital purpose, as it allows the organism tissue to be highly ordered with regard to the molecules it is built from. The second law of thermodynamics states that energy (and matter) tends to become more evenly spread out across the universe: to concentrate energy (or matter) in one specific place, it is necessary to spread out a greater amount of energy (as heat) across the remainder of the universe ("the surroundings"). Simpler organisms can achieve higher energy efficiencies than more complex ones, but the complex organisms can occupy ecological niches that are not available to their simpler brethren.

The conversion of a portion of the chemical energy to heat at each step in a metabolic pathway is the physical reason behind the pyramid of biomass observed in ecology: to take just the first step in the food chain, of the estimated 124.7 Pg/a of carbon that is fixed by photosynthesis, 64.3 Pg/a (52%) are used for the metabolism of green plants, i.e. reconverted into carbon dioxide and heat.

Bioenergetics

Bioenergetics is a field in biochemistry that concerns energy flow through living systems. This is an active area of biological research that includes the study of thousands of different cellular processes such as cellular respiration and the many other metabolic processes that can lead to production and utilization of energy in forms such as ATP molecules.

Bioenergetics is the part of biochemistry concerned with the energy involved in making and breaking of chemical bonds in the molecules found in biological organisms. It can also be defined as the study of energy relationships and energy transformations in living organisms.

Growth, development and metabolism are some of the central phenomena in the study of biological organisms. The role of energy is fundamental to such biological processes. The ability to harness energy from a variety of metabolic pathways is a property of all living organisms. Life is dependent on energy transformations; living organisms survive because of exchange of energy within and without.

In a living organism, chemical bonds are broken and made as part of the exchange and transformation of energy. Energy is available for work (such as mechanical work) or for other processes (such as chemical synthesis and anabolic processes in growth), when weak bonds are broken and stronger bonds are made. The production of stronger bonds allows release of usable energy.

Living organisms obtain energy from organic and inorganic materials. For example, lithotrophs can oxidize minerals such as nitrates or forms of sulfur, such as elemental sulfur, sulfites, and hydrogen sulfide to produce ATP. In photosynthesis, autotrophs can produce ATP using light energy. Heterotrophs must consume organic compounds. These are mostly carbohydrates, fats, and proteins. The amount of energy actually obtained by the organism is lower than the amount present in the food; there are losses in digestion, metabolism, and thermogenesis.

The materials are generally combined with oxygen to release energy, although some can also be oxidized anaerobically by various organisms. The bonds holding the molecules of nutrients together and the bonds holding molecules of free oxygen together are all relatively weak compared with the chemical bonds holding carbon dioxide and water together. The utilization of these materials is a form of slow combustion. That is why the energy content of food can be estimated with a bomb calorimeter. The materials are oxidized slowly enough that the organisms do not actually produce fire. The oxidation releases energy because stronger bonds have been formed. This net energy may evolve as heat, or some of which may be used by the organism for other purposes, such as breaking other bonds to do chemistry.

Living organisms produce ATP from energy sources via oxidative phosphorylation. The terminal phosphate bonds of ATP are relatively weak compared with the stronger bonds formed when ATP is broken down to adenosine monophosphate and phosphate and then dissolved in water. Here it is the energy of hydration that results in energy release. An organism's stockpile of ATP is used as a battery to store energy in cells, for intermediate metabolism. Utilization of chemical energy from such molecular bond rearrangement powers biological processes in every biological organism.

Types of Reactions

- Exergonic is a spontaneous reaction that releases energy. It is thermodynamically favoured. On the course of a reaction, energy needs to be put in, this activation energy drives the reactants from a stable state to a highly energetic unstable configuration. These reactants are usually complex molecules that are broken into simpler products. The entire reaction is usually catabolic. The release of energy (called Gibbs free energy) is negative and equal to - ΔG because energy is lost from the bonds formed by the products.
- Endergonic is an anabolic reaction that consumes energy. It has a positive ΔG because energy is required to break bonds.

  The free energy (ΔG) gained or lost in a reaction can be calculated: $\Delta G = \Delta H - T\Delta S$

  where G = Gibbs free energy, H = enthalpy, T = temperature, and S = entropy.

Cotransport

In August 1960, Robert K. Crane presented for the first time his discovery of the sodium-glucose cotransport as the mechanism for intestinal glucose absorption. Crane's discovery of cotransport was the first ever proposal of flux coupling in biology and was the most important event concerning carbohydrate absorption in the 20th century.

Chemiosmotic Theory

One of the major triumphs of bioenergetics is Peter D. Mitchell's chemiosmotic theory of how protons in aqueous solution function in the production of ATP in cell organelles such as mitochondria. This work earned Mitchell the 1978 Nobel Prize for Chemistry. Other cellular sources of ATP such as glycolysis were understood first, but such processes for direct coupling of enzyme activity to ATP production are not the major source of useful chemical energy in most cells. Chemiosmotic coupling is the major energy producing process in most cells, being utilized in chloroplasts and several single celled organisms in addition to mitochondria.

Cellular Respiration

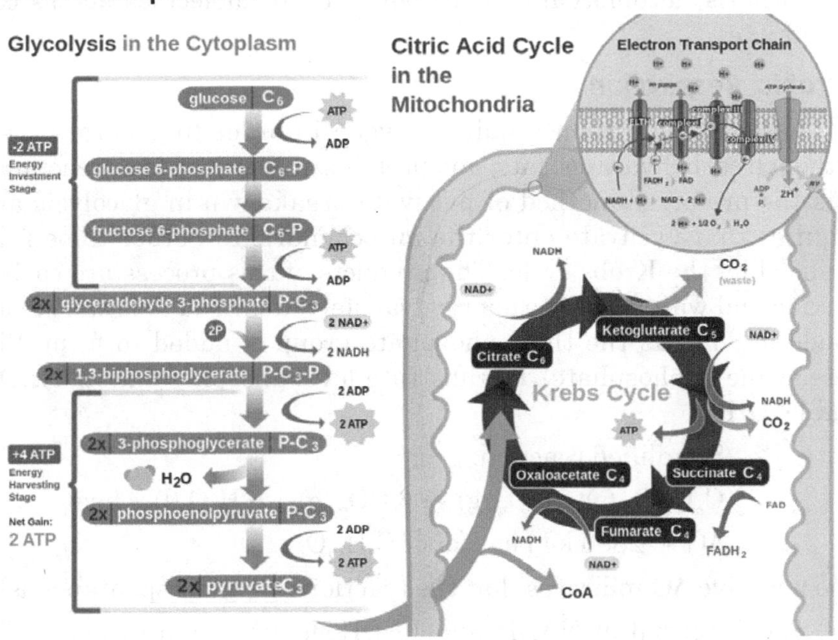

Figure: *Typical eukaryotic cell*

Cellular respiration is the set of metabolic reactions and processes that take place in the cells of organisms to convert biochemical energy from nutrients into adenosine triphosphate (ATP), and then release waste products. The reactions involved in respiration are catabolic reactions, which break large molecules into smaller ones, releasing energy in the process as weak so-called "high-energy" bonds are replaced by stronger bonds in the products. Respiration is one of the key ways a cell gains useful energy to fuel cellular activity. Cellular respiration is considered an exothermic redox reaction which releases heat. The overall reaction occurs in a series of biochemical steps, most of which are redox reactions themselves. Although technically, cellular respiration is a combustion reaction, it clearly does not resemble one when it occurs in a living cell due to slow release of energy from the series of reactions.

Nutrients that are commonly used by animal and plant cells in respiration include sugar, amino acids and fatty acids, and the most common oxidizing agent (electron acceptor) is molecular oxygen (O_2). The chemical energy stored in ATP (its third phosphate group is weakly bonded to the rest of the molecule and is cheaply broken allowing stronger bonds to form, thereby transferring energy for use by the cell) can then be used to drive processes requiring energy, including biosynthesis, locomotion or transportation of molecules across cell membranes.

Aerobic Respiration

Aerobic respiration requires oxygen in order to generate ATP. Although carbohydrates, fats, and proteins are consumed as reactants, it is the preferred method of pyruvate breakdown in glycolysis and requires that pyruvate enter the mitochondrion in order to be fully oxidized by the Krebs cycle. The products of this process are carbon dioxide and water, but the energy transferred is used to break strong bonds in ADP as the third phosphate group is added to form ATP (adenosine triphosphate), by substrate-level phosphorylation, NADH and $FADH_2$

Simplified reaction:

$$C_6H_{12}O_6\ (s) + 6\ O_2\ (g) \rightarrow 6\ CO_2\ (g) + 6\ H_2O\ (l) + \text{heat}$$

$$\Delta G = -2880 \text{ kJ per mol of } C_6H_{12}O_6$$

The negative ΔG indicates that the reaction can occur spontaneously.

The potential of NADH and $FADH_2$ is converted to more ATP through an electron transport chain with oxygen as the "terminal electron acceptor". Most of the ATP produced by aerobic cellular respiration is made by oxidative phosphorylation. This works by the energy released in the consumption of pyruvate being used to create a chemiosmotic potential by pumping protons across a membrane. This potential is then used to drive ATP synthase and produce ATP from ADP and a phosphate group. Biology textbooks often state that 38 ATP molecules can be made per oxidised glucose molecule during cellular respiration (2 from glycolysis, 2 from the Krebs cycle, and about 34 from the electron transport system). However, this maximum yield is never quite reached due to losses (leaky membranes) as well as the cost of moving pyruvate and ADP into the mitochondrial matrix, and current estimates range around 29 to 30 ATP per glucose.

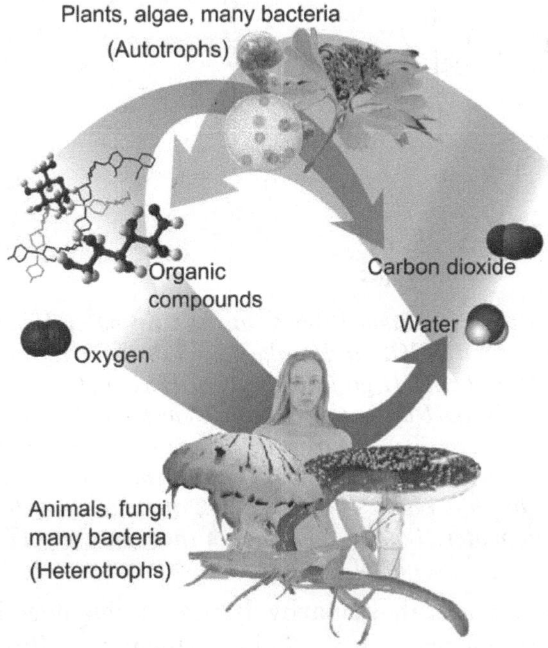

Figure: *Aerobic respirationis the main means by which both fungi and plants utilize chemical energy in the form of organic compounds that were previously created through photosynthesis.*

Aerobic metabolism is up to 15 times more efficient than anaerobic metabolism (which yields 2 molecules ATP per 1 molecule glucose). However some anaerobic organisms, such as methanogens are able to continue with anaerobic respiration, yielding more ATP by using other inorganic molecules (not oxygen) as final electron acceptors in the electron transport chain. They share the initial pathway of glycolysis but aerobic metabolism continues with the Krebs cycle and oxidative phosphorylation. The post-glycolytic reactions take place in the mitochondria in eukaryotic cells, and in the cytoplasm in prokaryotic cells.

Glycolysis

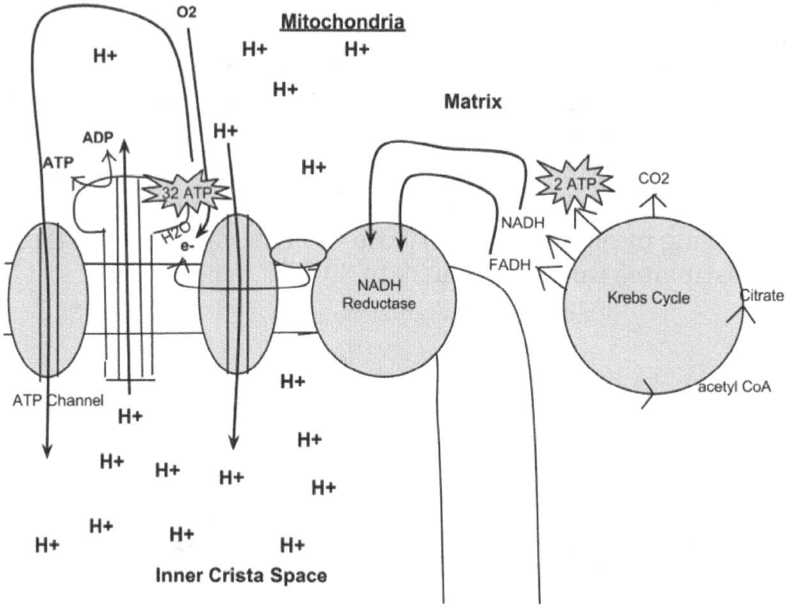

Figure: *Out of the cytoplasm it goes into the Krebs cycle with the acetyl CoA. It then mixes with CO2 and makes 2 ATP, NADH, and FADH. From there the NADH and FADH go into the NADH reductase, which produces the enzyme. The NADH pulls the enzyme's electrons to send through the electron transport chain. The electron transport chain pulls* H^+ *ions through the chain. From the electron transport chain, the released hydrogen ions make ADP for an end result of 32 ATP. 02 attracts itself to the left over electron to make water. Lastly, ATP leaves through the ATP channel and out of the mitochondria.*

Glycolysis is a metabolic pathway that takes place in the cytosol of cells in all living organisms. This pathway can function with or without the presence of oxygen. In humans, aerobic conditions produce pyruvate and anaerobic conditions produce lactate. In aerobic conditions, the process converts one molecule of glucose into two molecules of pyruvate (pyruvic acid), generating energy in the form of two net molecules of ATP. Four molecules of ATP per glucose are actually produced, however, two are consumed as part of the preparatory phase.

The initial phosphorylation of glucose is required to increase the reactivity (decrease its stability) in order for the molecule to be cleaved into two pyruvate molecules by the enzyme aldolase. During the pay-off phase of glycolysis, four phosphate groups are transferred to ADP by substrate-level phosphorylation to make four ATP, and two NADH are produced when the pyruvate are oxidized. The overall reaction can be expressed this way:

Glucose + 2 NAD^+ + 2 P_i + 2 ADP $\rightarrow$ 2 pyruvate + 2 NADH + 2 ATP + 2 H^+ + 2 H_2O + heat

Starting with glucose, 1 ATP is used to donate a phosphate to glucose to produce glucose 6-phosphate. Glycogen can be converted into glucose 6-phosphate as well with the help of glycogen phosphorylase.

During energy metabolism, glucose 6-phosphate becomes fructose 6-phosphate. An additional ATP is used to phosphorylate fructose 6-phosphate into fructose 1,6-disphosphate by the help of phosphofructokinase. Fructose 1,6-diphosphate then splits into two phosphorylated molecules with three carbon chains which later degrades into pyruvate.

Oxidative Decarboxylation of Pyruvate

Pyruvate is oxidized to acetyl-CoA and CO_2 by the pyruvate dehydrogenase complex (PDC). The PDC contains multiple copies of three enzymes and is located in the mitochondria of eukaryotic cells and in the cytosol of prokaryotes. In the conversion of pyruvate to acetyl-CoA, one molecule of NADH and one molecule of CO_2 is formed.

Citric Acid Cycle

This is also called the Krebs cycle or the tricarboxylic acid cycle. When oxygen is present, acetyl-CoA is produced from the pyruvate molecules created from glycolysis. When oxygen is present, the mitochondria will undergo aerobic respiration which leads to the Krebs cycle. However, if oxygen is not present, fermentation of the pyruvate molecule will occur. In the presence of oxygen, when acetyl-CoA is produced, the molecule then enters the citric acid cycle (Krebs cycle) inside the mitochondrial matrix, and gets oxidized to CO_2 while at the same time reducing NAD to NADH. NADH can be used by the electron transport chain to create further ATP as part of oxidative phosphorylation. To fully oxidize the equivalent of one glucose molecule, two acetyl-CoA must be metabolized by the Krebs cycle. Two waste products, H_2O and CO_2, are created during this cycle.

The citric acid cycle is an 8-step process involving different enzymes and co-enzymes. During the cycle, acetyl-CoA (2 carbons) + oxaloacetate (4 carbons) yields citrate (6 carbons), which is rearranged to a more reactive form called isocitrate (6 carbons). Isocitrate is modified to become α-ketoglutarate (5 carbons), succinyl-CoA, succinate, fumarate, malate, and, finally, oxaloacetate. The net gain of high-energy compounds from one cycle is 3 NADH, 1 $FADH_2$, and 1 GTP; the GTP may subsequently be used to produce ATP. Thus, the total yield from 1 glucose molecule (2 pyruvate molecules) is 6 NADH, 2 $FADH_2$, and 2 ATP.

Oxidative Phosphorylation

In eukaryotes, oxidative phosphorylation occurs in the mitochondrial cristae. It comprises the electron transport chain that establishes a proton gradient (chemiosmotic potential) across the inner membrane by oxidizing the NADH produced from the Krebs cycle. ATP is synthesised by the ATP synthase enzyme when the chemiosmotic gradient is used to drive the phosphorylation of ADP. The electrons are finally transferred to exogenous oxygen and, with the addition of two protons, water is formed.

Efficiency of ATP Production

The table below describes the reactions involved when one glucose molecule is fully oxidized into carbon dioxide. It is assumed that all the reduced coenzymes are oxidized by the electron transport chain and used for oxidative phosphorylation.

Step	*coenzyme yield*	*ATP yield*	*Source of ATP*
Glycolysis preparatory phase		? 2	Phosphorylation of glucose and fructose 6-phosphate uses two ATP from the cytoplasm.
Glycolysis pay-off phase		4	Substrate-level phosphorylation
	2 NADH	3	Oxidative phosphorylation – Each NADH produces net 1.5 ATP (instead of usual 2.5) due to NADH transport over the mitochondrial membrane
Oxidative decarboxylation of pyruvate	2 NADH	5	Oxidative phosphorylation
Krebs cycle		2	Substrate-level phosphorylation
	6 NADH	15	Oxidative phosphorylation
	2 $FADH_2$	3	Oxidative phosphorylation
Total yield		30 ATP	From the complete oxidation of one glucose molecule to carbon dioxide and oxidation of all the reduced coenzymes.

Although there is a theoretical yield of 38 ATP molecules per glucose during cellular respiration, such conditions are generally not realized due to losses such as the cost of moving pyruvate (from glycolysis), phosphate, and ADP (substrates for ATP synthesis) into the mitochondria. All are actively transported using carriers that utilize the stored energy in the proton electrochemical gradient.

- Pyruvate is taken up by a specific, low Km transporter to bring it into the mitochondrial matrix for oxidation by the pyruvate dehydrogenase complex.
- The phosphate carrier (PiC) mediates the electroneutral exchange (antiport) of phosphate ($H_2PO_4^-$; P_i) for OH^- or symport of phosphate and protons (H^+) across the inner membrane, and the driving force for moving phosphate ions into the mitochondria is the proton motive force.
- The ATP-ADP translocase (also called adenine nucleotide translocase, ANT) is an antiporter and exchanges ADP and ATP across the inner membrane. The driving force is due to the ATP (-4) having a more negative charge than the ADP (–3), and thus it dissipates some of the electrical component of the proton electrochemical gradient.

The outcome of these transport processes using the proton electrochemical gradient is that more than 3 H^+ are needed to make 1 ATP. Obviously this reduces the theoretical efficiency of the whole process and the likely maximum is closer to 28–30 ATP molecules. In practice the efficiency may be even lower due to the inner membrane of the mitochondria being slightly leaky to protons. Other factors may also dissipate the proton gradient creating an apparently leaky mitochondria. An uncoupling protein known as thermogenin is expressed in some cell types and is a channel that can transport protons. When this protein is active in the inner membrane it short circuits the coupling between the electron transport chain and ATP synthesis. The potential energy from the proton gradient is not used to make ATP but generates heat. This is particularly important in brown fat thermogenesis of newborn and hibernating mammals.

According to some of newer sources the ATP yield during aerobic respiration is not 36–38, but only about 30–32 ATP molecules / 1 molecule of glucose , because:

- ATP : NADH+H^+ and ATP : $FADH_2$ ratios during the oxidative phosphorylation appear to be not 3 and 2, but 2.5 and 1.5 respectively. Unlike in the substrate-level phosphorylation, the stoichiometry here is difficult to establish.
    - o ATP synthase produces 1 ATP / 3 H^+. However the exchange of matrix ATP for cytosolic ADP and Pi (antiport with OH^- or symport with H^+) mediated by ATP–ADP translocase and phosphate carrier consumes 1 H^+ / 1 ATP due to regeneration of the transmembrane potential changed during this transfer, so the net ratio is 1 ATP : 4 H^+.

- o The mitochondrial electron transport chain proton pump transfers across the inner membrane 10 H^+ / 1 NADH+H^+ (4 + 2 + 4) or 6 H^+ / 1 $FADH_2$ (2 + 4).

So the final stoichiometry is

1 NADH+H^+ : 10 H^+ : 10/4 ATP = 1 NADH+H^+ : 2.5 ATP

1 $FADH_2$: 6 H^+ : 6/4 ATP = 1 $FADH_2$: 1.5 ATP

- ATP : NADH+H^+ coming from glycolysis ratio during the oxidative phosphorylation is
  - o 1.5, as for $FADH_2$, if hydrogen atoms ($2H^+$+$2e^-$) are transferred from cytosolic NADH+H^+ to mitochondrial FAD by the glycerol phosphate shuttle located in the inner mitochondrial membrane.
  - o 2.5 in case of malate-aspartate shuttle transferring hydrogen atoms from cytosolic NADH+H^+ to mitochondrial NAD^+

So finally we have, per molecule of glucose

- Substrate-level phosphorylation: 2 ATP from glycolysis + 2 ATP (directly GTP) from Krebs cycle
- Oxidative phosphorylation
  - o 2 NADH+H^+ from glycolysis: 2 × 1.5 ATP (if glycerol phosphate shuttle transfers hydrogen atoms) or 2 × 2.5 ATP (malate-aspartate shuttle)
  - o 2 NADH+H^+ from the oxidative decarboxylation of pyruvate and 6 from Krebs cycle: 8 × 2.5 ATP
  - o 2 $FADH_2$ from the Krebs cycle: 2 × 1.5 ATP

Altogether this gives 4 + 3 (or 5) + 20 + 3 = 30 (or 32) ATP per molecule of glucose

The total ATP yield in ethanol or lactic acid fermentation is only 2 molecules coming from glycolysis, because pyruvate is not transferred to the mitochondrion and finally oxidized to the carbon dioxide (CO_2), but reduced to ethanol or lactic acid in the cytoplasm.

Fermentation

Without oxygen, pyruvate (pyruvic acid) is not metabolized by cellular respiration but undergoes a process of fermentation. The pyruvate is not transported into the mitochondrion, but remains in the cytoplasm, where it is converted to waste products that may be removed from the cell. This serves the purpose of oxidizing the electron carriers so that they can perform glycolysis again and removing the excess pyruvate. Fermentation oxidizes NADH to NAD+ so it can be re-used in glycolysis. In the absence of oxygen, fermentation prevents the buildup of NADH in the cytoplasm and provides NAD+ for glycolysis. This waste product varies depending on the organism. In skeletal muscles, the waste product is lactic acid.

This type of fermentation is called lactic acid fermentation. In strenuous exercise, when energy demands exceed energy supply, the respiratory chain cannot process all of the hydrogen atoms joined by NADH. During anaerobic glycolysis, NAD+ regenerates when pairs of hydrogen combine with pyruvate to form lactate. Lactate formation is catalyzed by lactate dehydrogenase in a reversible reaction. Lactate can also be used as an indirect precursor for liver glycogen. During recovery, when oxygen becomes available, NAD+ attaches to hydrogen from lactate to form ATP. In yeast, the waste products are ethanol and carbon dioxide. This type of fermentation is known as alcoholic or ethanol fermentation. The ATP generated in this process is made by substrate-level phosphorylation, which does not require oxygen.

Fermentation is less efficient at using the energy from glucose: only 2 ATP are produced per glucose, compared to the 38 ATP per glucose nominally produced by aerobic respiration. This is because the waste products of fermentation still contain chemical potential energy that can be released by oxidation. Ethanol, for example, can be burned in an internal combustion engine like gasoline. Glycolytic ATP, however, is created more quickly. For prokaryotes to continue a rapid growth rate when they are shifted from an aerobic environment to an anaerobic environment, they must increase the rate of the glycolytic reactions. For multicellular organisms, during short bursts of strenuous activity, muscle cells use fermentation to supplement the ATP production from the slower aerobic respiration, so fermentation may be used by a cell even before the oxygen levels are depleted, as is the case in sports that do not require athletes to pace themselves, such as sprinting.

Anaerobic Respiration

Cellular respiration is the process by which biological fuels are oxidised in the presence of an inorganic electron acceptor (such as oxygen) to produce large amounts of energy, to drive the bulk production of ATP.

Anaerobic respiration is used by some microorganisms in which neither oxygen (aerobic respiration) nor pyruvate derivatives (fermentation) is the final electron acceptor. Rather, an inorganic acceptor such as sulfate or nitrate is used. Such organisms are typically found in unusual places such as underwater caves or near the lava shoots at the bottom of the ocean.

Many high-school biology textbooks incorrectly refer to fermentation (e.g., to lactate) as anaerobic respiration.

Liposome

Liposomes are composite structures made of phospholipids and may contain small amounts of other molecules. Though liposomes can vary in size from low micrometer range to tens of micrometers, unilamellar liposomes, as pictured here, are typically in the lower size range with various targeting ligands attached to their surface allowing for their surface-attachment and accumulation in pathological areas for treatment of disease.

A liposome is an artificially-prepared spherical vesicle composed of a lamellar phase lipid bilayer. The liposome can be used as a vehicle for administration of nutrients and pharmaceutical drugs. Liposomes can be prepared by disrupting biological membranes (such as by sonication).

Liposomes are often composed of phosphatidylcholine-enriched phospholipids and may also contain mixed lipid chains with surfactant properties such as egg phosphatidylethanolamine. A liposome design may employ surface ligands for attaching to unhealthy tissue.

The major types of liposomes are the multilamellar vesicle (MLV), the small unilamellar liposome vesicle (SUV), the large unilamellar vesicle (LUV), and the cochleate vesicle.

Liposomes should not be confused with micelles and reverse micelles composed of monolayers.

Discovery

The word *liposome* derives from two Greek words: *lipo* ("fat") and *soma* ("body"); it is so named because its composition is primarily of phospholipid.

Liposomes were first described by British haematologist Alec D Bangham in 1961 (published 1964), at the Babraham Institute, in Cambridge. They were discovered when Bangham and R. W. Horne were testing the institute's new electron microscope by adding negative stain to dry phospholipids. The resemblance to the plasmalemma was obvious, and the microscope pictures served as the first evidence for the cell membrane being a bilayer lipid structure. Their integrity as a closed, bilayer structure, that could release its contents after detergent treatment (structure-linked latency) was established by Bangham, Standish and Weissmann in the next year. Weissmann - during a Cambridge pub discussion with Bangham - first named the structures "liposomes" after the lysosome, which his laboratory had been studying: a simple organelle the structure-linked latency of which could be disrupted by detergents and streptolysins. Liposomes can be easily distinguished from micelles and hexagonal lipid phases by negative staining transmission electron microscopy.

Alec Douglas Bangham with colleagues Jeff Watkins and Malcolm Standish wrote the 1965 paper that effectively launched the liposome "industry". Around this time he was joined at Babraham by Gerald Weissmann, an American physician with an interest in lysosomes. Now an emeritus professor at New York University School of Medicine, Weissmann recalls the two of them sitting in a Cambridge pub and reflecting on the role of lipid sheets in separating the interior of the cell from the exterior milieu. This insight, they felt, was to cell function what the discovery of the double helix had been to genetics. Bangham had called his lipid structures "multilamellar smectic mesophases" or sometimes "Banghasomes". It was Weissmann who proposed the more user-friendly term liposome.

Mechanism

A liposome encapsulates a region of aqueous solution inside a hydrophobic membrane; dissolved hydrophilic solutes cannot readily pass through the lipids. Hydrophobic chemicals can be dissolved into the membrane, and in this way liposome can carry both hydrophobic molecules and hydrophilic molecules. To deliver the molecules to sites of action, the lipid bilayer can fuse with other bilayers such as the cell membrane, thus delivering the liposome contents. By making liposomes in a solution of DNA or drugs (which would normally be unable to diffuse through the membrane) they can be (indiscriminately) delivered past the lipid bilayer. A liposome does not necessarily have lipophobic contents, such as water, although it usually does.

Liposomes are used as models for artificial cells. Liposomes can also be designed to deliver drugs in other ways. Liposomes that contain low (or high) pH can be constructed such that dissolved aqueous drugs will be charged in solution (i.e., the pH is outside the drug's pI range). As the pH naturally neutralizes within the liposome (protons can pass through some membranes), the drug will also be neutralized, allowing it to freely pass through a membrane. These liposomes work to deliver drug by diffusion rather than by direct cell fusion.

A similar approach can be exploited in the biodetoxification of drugs by injecting empty liposomes with a transmembrane pH gradient. In this case the vesicles act as sinks to scavenge the drug in the blood circulation and prevent its toxic effect. Another strategy for liposome drug delivery is to target endocytosis events. Liposomes can be made in a particular size range that makes them viable targets for natural macrophage phagocytosis. These liposomes may be digested while in the macrophage's phagosome, thus releasing its drug. Liposomes can also be decorated with opsonins and ligands to activate endocytosis in other cell types.

The use of liposomes for transformation or transfection of DNA into a host cell is known as lipofection.

In addition to gene and drug delivery applications, liposomes can be used as carriers for the delivery of dyes to textiles, pesticides to plants, enzymes and nutritional supplements to foods, and cosmetics to the skin.

Liposomes are also used as outer shells of some microbubble contrast agents used in contrast-enhanced ultrasound.

Dietary and Nutritional Supplements

Regarding the use of liposomes as a carrier of dietary and nutritional supplements; until very recently the use of liposomes were primarily directed at targeted drug delivery. However, the versatile abilities of liposomes are now being discovered in other settings. Liposomes are presently being cleverly implemented for the specific oral delivery of certain dietary and nutritional supplements.

A very small number of dietary and nutritional supplement companies are currently pioneering the benefits of this unique science towards this new application. This new direction and employment of liposome science is in part due to the low absorption and bioavailability rates of traditional oral dietary and nutritional tablets and capsules. The low oral bioavailability and absorption of many nutrients is clinically well documented. Therefore the natural encapsulation of lypophilic and hydrophilic nutrients within liposomes has made for a very effective method of bypassing the destructive elements of the gastric system and aiding the encapsulated nutrient to be delivered to the cells and tissues.

It is important to note that certain influential factors have far reaching effects on the percentage of liposome that are yielded in manufacturing. These influences also have an effect on the actual amount of realized liposome entrapment and the actual quality of the liposomes themselves. These are very crucial elements which lead to the long term stability of the liposomes. These complex yet significant factors are the following:

(1) The actual manufacturing method and preparation of the liposomes themselves;
(2) The constitution, quality, and type of raw phospholipid used in the formulation and manufacturing of the liposomes;
(3) The ability to create homogeneous liposome particle sizes that are stable and hold their encapsulated payload.

These primary and key elements comprise the foundation of an effective liposome carrier for use in increasing the bioavailability of oral dosages of dietary and nutritional supplements.

Manufacturing

The correct choice of liposome preparation method depends on the following parameters:

1. the physicochemical characteristics of the material to be entrapped and those of the liposomal ingredients;
2. the nature of the medium in which the lipid vesicles are dispersed
3. the effective concentration of the entrapped substance and its potential toxicity;
4. additional processes involved during application/delivery of the vesicles;
5. optimum size, polydispersity and shelf-life of the vesicles for the intended application; and,
6. batch-to-batch reproducibility and possibility of large-scale production of safe and efficient liposomal products

Formation of liposomes and nanoliposomes is not a spontaneous process. Lipid vesicles are formed when phospholipids such as lecithin are placed in water and consequently form one bilayer or a series of bilayers, each separated by water molecules, once enough energy is supplied.

Liposomes can be created by sonicating phosphatidylcholine rich phospholipids in water. Low shear rates create multilamellar liposomes, which have many layers like an onion. Continued high-shear sonication tends to form smaller unilamellar liposomes. In this technique, the liposome contents are the same as the contents of the aqueous phase. Sonication is generally considered a "gross" method of preparation as it can damage the structure of the drug to be encapsulated. Newer methods such as extrusion and Mozafari method are employed to produce materials for human use.

Prospect

Further advances in liposome research have been able to allow liposomes to avoid detection by the body's immune system, specifically, the cells of reticuloendothelial system (RES). These liposomes are known as "stealth liposomes", and are constructed with PEG (Polyethylene Glycol) studding the outside of the membrane. The PEG coating, which is inert in the body, allows for longer circulatory life for the drug delivery mechanism. However, research currently seeks to investigate at what amount of PEG coating the PEG actually hinders binding of the liposome to the delivery site.

In addition to a PEG coating, most stealth liposomes also have some sort of biological species attached as a ligand to the liposome in order to enable binding via a specific expression on the targeted drug delivery site. These targeting ligands could be monoclonal antibodies (making an immunoliposome), vitamins, or specific antigens. Targeted liposomes can target nearly any cell type in the body and deliver drugs that would naturally be systemically delivered. Naturally toxic drugs can be much less toxic if delivered only to diseased tissues. Polymersomes, morphologically related to liposomes, can also be used this way.

In case of tumor development, certain anticancer drugs such as doxorubicin (Doxil) and daunorubicin are provided through liposomes. Liposomal cisplatin has received orphan drug designation for pancreatic cancer from EMEA.

Lamella

A lamella (plural, lamellae), in cell biology, is used to describe numerous plate or disc-like structures at both a tissue and cellular level.

An example of this begin an extension of a thylakoid within a chloroplast, linking a thylakoid within one granum to one in another. They are the sites of photosystem I. Simply put, lamellae may be considered as a pair of membranes containing chlorophyll.

Chloroplasts are characterized by a system of membranes embedded in a hydrophobic proteinaceous matrix, or stroma. The basic unit of the membrane system is a flattened single vesicle called the thylakoid; thylakoids stack into grana (sing, granum). All the thylakoids of a granum are connected with each other, and the grana are connected by intergranal lamellae.

It is placed between the two primary cell walls of two plant cells and made up of intracellular matrix. The lamella comprises a mixture of polygalacturons (D-galacturonic acid) and neutral carbohydrates. It is soluble in the pectinase enzyme.

A lamella, in cell biology, is also used to describe the leading edge of a motile cell, of which the lamellipodia is the most forward portion.

The lipid bilayer core of biological membranes, is also called lamellar phase. Thus, each bilayer of multilamellar liposomes and wall of a unilamellar liposome is also referred to as *lamella*.

Adenosine Triphosphate

Adenosine triphosphate (ATP) is a nucleoside triphosphate used in cells as a coenzyme, often called the "molecular unit of currency" of intracellular energy transfer.

ATP transports chemical energy within cells for metabolism. It is one of the end products of photophosphorylation, cellular respiration, and fermentation and used by enzymes and structural proteins in many cellular processes, including biosynthetic reactions, motility, and cell division. One molecule of ATP contains three phosphate groups, and it is produced by a wide variety of enzymes, including ATP synthase, from adenosine diphosphate (ADP) or adenosine monophosphate (AMP) and various phosphate group donors. Substrate-level phosphorylation, oxidative phosphorylation in cellular respiration, and photophosphorylation in photosynthesis are three major mechanisms of ATP biosynthesis.

Metabolic processes that use ATP as an energy source convert it back into its precursors. ATP is therefore continuously recycled in organisms: the human body, which on average contains only 250 grams (8.8 oz) of ATP, turns over its own body weight equivalent in ATP each day.

ATP is used as a substrate in signal transduction pathways by kinases that phosphorylate proteins and lipids. It is also used by adenylate cyclase, which uses ATP to produce the second messenger molecule cyclic AMP. The ratio between ATP and AMP is used as a way for a cell to sense how much energy is available and control the metabolic pathways that produce and consume ATP. Apart from its roles in signalling and energy metabolism, ATP is also incorporated into nucleic acids by polymerases in the process of transcription. ATP is the neurotransmitter believed to signal the sense of taste.

The structure of this molecule consists of a purine base (adenine) attached to the 1' carbon atom of a pentose sugar (ribose). Three phosphate groups are attached at the 5' carbon atom of the pentose sugar. It is the addition and removal of these phosphate groups that inter-convert ATP, ADP and AMP. When ATP is used in DNA synthesis, the ribose sugar is first converted to deoxyribose by ribonucleotide reductase.

ATP was discovered in 1929 by Karl Lohmann, Fiske and Y. Subbarow of Harvard Medical School but its correct structure was not determined until some years later. It was proposed to be the main energy transfer molecule in the cell by Fritz Albert Lipmann in 1941, that is, being the intermediary molecule between energy-yielding (exergonic) and energy-requiring (endergonic) reactions. It was first artificially synthesized by Alexander Todd in 1948.

Physical and Chemical Properties

ATP consists of adenosine — composed of an adenine ring and a ribose sugar — and three phosphate groups (triphosphate). The phosphoryl groups, starting with the group closest to the ribose, are referred to as the alpha (α), beta (β), and gamma (γ) phosphates. Consequently, it is closely related to the adenosine nucleotide, a monomer of RNA. ATP is highly soluble in water and is quite stable in solutions between pH 6.8 and 7.4, but is rapidly hydrolysed at extreme pH. Consequently, ATP is best stored as an anhydrous salt.

ATP is an unstable molecule in unbuffered water, in which it hydrolyses to ADP and phosphate. This is because the strength of the bonds between the phosphate groups in ATP is less than the strength of the hydrogen bonds (hydration bonds), between its products (ADP + phosphate), and water. Thus, if ATP and ADP are in chemical equilibrium in water, almost all of the ATP will eventually be converted to ADP. A system that is far from equilibrium contains Gibbs free energy, and is capable of doing work. Living cells maintain the ratio of ATP to ADP at a point ten orders of magnitude from equilibrium, with ATP concentrations fivefold higher than the concentration of ADP. This displacement from equilibrium means that the hydrolysis of ATP in the cell releases a large amount of free energy.

Two phosphoanhydride bonds (those that connect adjacent phosphates) in an ATP molecule are responsible for the high energy content of this molecule. In the context of biochemical reactions, these anhydride bonds are frequently—and sometimes controversially—referred to as *high-energy bonds* (despite the fact it takes energy to break bonds). Energy stored in ATP may be released upon hydrolysis of the anhydride bonds. The primary phosphate group on the ATP molecule that is hydrolyzed when energy is needed to drive anabolic reactions is the γ-phosphate group. Located the farthest from the ribose sugar, it has a higher energy of hydrolysis than either the α – or β – phosphate. The bonds formed after hydrolysis—or the phosphorylation of a residue by ATP—are lower in energy than the phosphoanhydride bonds of ATP. During enzyme-catalyzed hydrolysis of ATP or phosphorylation by ATP, the available free energy can be harnessed by a living system to do work.

Any unstable system of potentially reactive molecules could potentially serve as a way of storing free energy, if the cell maintained their concentration far from the equilibrium point of the reaction. However, as is the case with most polymeric biomolecules, the breakdown of RNA, DNA, and ATP into simpler monomers is driven by both energy-release and entropy-increase considerations, in both standard concentrations, and also those concentrations encountered within the cell.

Biosynthesis

The ATP concentration inside the cell is typically 1–10 mM. ATP can be produced by redox reactions using simple and complex sugars (carbohydrates) or lipids as an energy source. For complex fuels to be synthesized into ATP, they first need to be broken down into smaller, more simple molecules. Carbohydrates are hydrolysed into simple sugars, such as glucose and fructose. Fats (triglycerides) are metabolised to give fatty acids and glycerol.

The overall process of oxidizing glucose to carbon dioxide is known as cellular respiration and can produce about 30 molecules of ATP from a single molecule of glucose. ATP can be produced by a number of distinct cellular processes; the three main pathways used to generate energy in eukaryotic organisms are glycolysis and the citric acid cycle/ oxidative phosphorylation, both components of cellular respiration; and beta-oxidation. The majority of this ATP production by a non-photosynthetic aerobic eukaryote takes place in the mitochondria, which can make up nearly 25% of the total volume of a typical cell.

Glycolysis

In glycolysis, glucose and glycerol are metabolized to pyruvate via the glycolytic pathway. In most organisms, this process occurs in the cytosol, but, in some protozoa such as the kinetoplastids, this is carried out in a specialized organelle called the glycosome. Glycolysis generates a net two molecules of ATP through substrate phosphorylation catalyzed by two enzymes: PGK and pyruvate kinase. Two molecules of NADH are also produced, which can be oxidized via the electron transport chain and result in the generation of additional ATP by ATP synthase. The pyruvate generated as an end-product of glycolysis is a substrate for the Krebs Cycle.

Glucose

In the mitochondrion, pyruvate is oxidized by the pyruvate dehydrogenase complex to Acetyl group, which is fully oxidized to carbon dioxide by the citric acid cycle (also known as the Krebs Cycle). Every "turn" of the citric acid cycle produces two molecules of carbon dioxide, one molecule of the ATP equivalent guanosine triphosphate (GTP) through substrate-level phosphorylation catalyzed by succinyl-CoA synthetase, three molecules of the reduced coenzyme NADH, and one molecule of the reduced coenzyme $FADH_2$. Both of these latter molecules are recycled to their oxidized states (NAD^+ and FAD, respectively) via the electron transport chain, which generates additional ATP by oxidative phosphorylation.

The oxidation of an NADH molecule results in the synthesis of 2–3 ATP molecules, and the oxidation of one $FADH_2$ yields between 1–2 ATP molecules. The majority of cellular ATP is generated by this process. Although the citric acid cycle itself does not involve molecular oxygen, it is an obligately aerobic process because O_2 is needed to recycle the reduced NADH and $FADH_2$ to their oxidized states. In the absence of oxygen the citric acid cycle will cease to function due to the lack of available NAD^+ and FAD.

The generation of ATP by the mitochondrion from cytosolic NADH relies on the malate-aspartate shuttle (and to a lesser extent, the glycerol-phosphate shuttle) because the inner mitochondrial membrane is impermeable to NADH and NAD^+. Instead of transferring the generated NADH, a malate dehydrogenase enzyme converts oxaloacetate to malate, which is translocated to the mitochondrial matrix. Another malate dehydrogenase-catalyzed reaction occurs in the opposite direction, producing oxaloacetate and NADH from the newly transported malate and the mitochondrion's interior store of NAD^+. A transaminase converts the oxaloacetate to aspartate for transport back across the membrane and into the intermembrane space.

In oxidative phosphorylation, the passage of electrons from NADH and $FADH_2$ through the electron transport chain powers the pumping of protons out of the mitochondrial matrix and into the intermembrane space. This creates a proton motive force that is the net effect of a pH gradient and an electric potential gradient across the inner mitochondrial membrane. Flow of protons down this potential gradient — that is, from the intermembrane space to the matrix — provides the driving force for ATP synthesis by ATP synthase. This enzyme contains a rotor subunit that physically rotates relative to the static portions of the protein during ATP synthesis.

Most of the ATP synthesized in the mitochondria will be used for cellular processes in the cytosol; thus it must be exported from its site of synthesis in the mitochondrial matrix. The inner membrane contains an antiporter, the ADP/ATP translocase, which is an integral membrane protein used to exchange newly synthesized ATP in the matrix for ADP in the intermembrane space. This translocase is driven by the membrane potential, as it results in the movement of about 4 negative charges out of the mitochondrial membrane in exchange for 3 negative charges moved inside. However, it is also necessary to transport phosphate into the mitochondrion; the phosphate carrier moves a proton in with each phosphate, partially dissipating the proton gradient.

Beta Oxidation

Fatty acids can also be broken down to acetyl-CoA by beta-oxidation. Each round of this cycle reduces the length of the acyl chain by two carbon atoms and produces one NADH and one $FADH_2$ molecule, which are used to generate ATP by oxidative phosphorylation. Because NADH and $FADH_2$ are energy-rich molecules, dozens of ATP molecules can be generated by the beta-oxidation of a single long acyl chain. The high energy yield of this process and the compact storage of fat explain why it is the most dense source of dietary calories.

Fermentation

Fermentation entails the generation of energy via the process of substrate-level phosphorylation in the absence of a respiratory electron transport chain. In most eukaryotes, glucose is used as both an energy store and an electron donor. The equation for the oxidation of glucose to lactic acid is:

$$C_6H_{12}O_6 \rightarrow 2CH_3CH(OH)COOH + 2ATP$$

Anaerobic Respiration

Anaerobic respiration is the process of respiration using an electron acceptor other than O_2. In prokaryotes, multiple electron acceptors can be used in anaerobic respiration. These include nitrate, sulphate or carbon dioxide.

These processes lead to the ecologically important processes of denitrification, sulphate reduction and acetogenesis, respectively.

ATP Replenishment by Nucleoside Diphosphate Kinases

ATP can also be synthesized through several so-called "replenishment" reactions catalyzed by the enzyme families of nucleoside diphosphate kinases (NDKs), which use other nucleoside triphosphates as a high-energy phosphate donor, and the ATP: guanido-phosphotransferase family,

ATP Production During Photosynthesis

In plants, ATP is synthesized in thylakoid membrane of the chloroplast during the light-dependent reactions of photosynthesis in a process called photophosphorylation. Here, light energy is used to pump protons across the chloroplast membrane. This produces a proton-motive force and this drives the ATP synthase, exactly as in oxidative phosphorylation. Some of the ATP produced in the chloroplasts is consumed in the Calvin cycle, which produces triose sugars.

ATP Recycling

The total quantity of ATP in the human body is about 0.2 mole. The majority of ATP is not usually synthesised *de novo*, but is generated from ADP by the aforementioned processes. Thus, at any given time, the total amount of ATP + ADP remains fairly constant.

The energy used by human cells requires the hydrolysis of 100 to 150 moles of ATP daily, which is around 50 to 75 kg. A human will typically use up his or her body weight of ATP over the course of the day. This means that each ATP molecule is recycled 500 to 750 times during a single day (100 / 0.2 = 500). ATP cannot be stored, hence its consumption closely follows its synthesis. However a total of around 5g of ATP is used by cell processes at any time in the body.

Regulation of Biosynthesis

ATP production in an aerobic eukaryotic cell is tightly regulated by allosteric mechanisms, by feedback effects, and by the substrate concentration dependence of individual enzymes within the glycolysis and oxidative phosphorylation pathways. Key control points occur in enzymatic reactions that are so energetically favourable that they are effectively irreversible under physiological conditions.

In glycolysis, hexokinase is directly inhibited by its product, glucose-6-phosphate, and pyruvate kinase is inhibited by ATP itself.

The main control point for the glycolytic pathway is phosphofructokinase (PFK), which is allosterically inhibited by high concentrations of ATP and activated by high concentrations of AMP.

The inhibition of PFK by ATP is unusual, since ATP is also a substrate in the reaction catalyzed by PFK; the biologically active form of the enzyme is a tetramer that exists in two possible conformations, only one of which binds the second substrate fructose-6-phosphate (F6P).

The protein has two binding sites for ATP — the active site is accessible in either protein conformation, but ATP binding to the inhibitor site stabilizes the conformation that binds F6P poorly. A number of other small molecules can compensate for the ATP-induced shift in equilibrium conformation and reactivate PFK, including cyclic AMP, ammonium ions, inorganic phosphate, and fructose 1,6 and 2,6 biphosphate.

The citric acid cycle is regulated mainly by the availability of key substrates, particularly the ratio of NAD^+ to NADH and the concentrations of calcium, inorganic phosphate, ATP, ADP, and AMP. Citrate — the molecule that gives its name to the cycle — is a feedback inhibitor of citrate synthase and also inhibits PFK, providing a direct link between the regulation of the citric acid cycle and glycolysis.

Functions in Cells

Metabolism, Synthesis, and Active Transport

ATP is consumed in the cell by energy-requiring (endothermic) processes and can be generated by energy-releasing (exothermic) processes. In this way ATP transfers energy between spatially separate metabolic reactions. ATP is the main energy source for the majority of cellular functions. This includes the synthesis of macromolecules, including DNA and RNA, and proteins. ATP also plays a critical role in the transport of macromolecules across cell membranes, e.g. exocytosis and endocytosis.

Roles in Cell structure and Locomotion

ATP is critically involved in maintaining cell structure by facilitating assembly and disassembly of elements of the cytoskeleton. In a related process, ATP is required for the shortening of actin and myosin filament crossbridges required for muscle contraction. This latter process is one of the main energy requirements of animals and is essential for locomotion and respiration.

Cell Signalling

Extracellular Signalling

ATP is also a signalling molecule. ATP, ADP, or adenosine are recognised by purinergic receptors. Purinoreceptors might be the most abundant receptors in mammalian tissues.

In humans, this signalling role is important in both the central and peripheral nervous system. Activity-dependent release of ATP from synapses, axons and glia activates purinergic membrane receptors known as P2. The *P2Y* receptors are *metabotropic*, i.e. G protein-coupled and modulate mainly intracellular calcium and sometimes cyclic AMP levels. Though named between $P2Y_1$ and $P2Y_{15}$, only nine members of the P2Y family have been cloned, and some are only related through weak homology and several ($P2Y_5$, $P2Y_7$, $P2Y_9$, $P2Y_{10}$) do not function as receptors that raise cytosolic calcium. The *P2X ionotropic* receptor subgroup comprises seven members ($P2X_1$–$P2X_7$), which are ligand-gated Ca^{2+}-permeable ion channels that open when bound to an extracellular purine nucleotide.

In contrast to P2 receptors (agonist order ATP > ADP > AMP > ADO), purinergic nucleoside triphosphates like ATP are not strong agonists of P1 receptors, which are strongly activated by adenosine and other nucleosides (ADO > AMP > ADP > ATP). P1 receptors have A1, A2a, A2b, and A3 subtypes ("A" as a remnant of old nomenclature of *adenosine receptor*), all of which are G protein-coupled receptors, A1 and A3 being coupled to Gi, and A2a and A2b being coupled to Gs. All adenosine receptors were shown to activate at least one subfamily of mitogen-activated protein kinases. The actions of adenosine are often antagonistic or synergistic to the actions of ATP. In the CNS, adenosine has multiple functions, such as modulation of neural development, neuron and glial signalling and the control of innate and adaptive immune systems.

Intracellular Signalling

ATP is critical in signal transduction processes. It is used by kinases as the source of phosphate groups in their phosphate transfer reactions. Kinase activity on substrates such as proteins or membrane lipids are a common form of signal transduction. Phosphorylation of a protein by a kinase can activate this cascade such as the mitogen-activated protein kinase cascade.

ATP is also used by adenylate cyclase and is transformed to the second messenger molecule cyclic AMP, which is involved in triggering calcium signals by the release of calcium from intracellular stores. This form of signal transduction is particularly important in brain function, although it is involved in the regulation of a multitude of other cellular processes.

DNA and RNA Synthesis

In all known organisms, the Deoxyribonucleotides that make up DNA are synthesized by the action of ribonucleotide reductase (RNR) enzymes on their corresponding ribonucleotides.

These enzymes reduce the sugar residue from ribose to deoxyribose by removing oxygen from the 2' hydroxyl group; the substrates are ribonucleoside diphosphates and the products deoxyribonucleoside diphosphates (the latter are denoted dADP, dCDP, dGDP, and dUDP respectively.)

All ribonucleotide reductase enzymes use a common sulfhydryl radical mechanism reliant on reactive cysteine residues that oxidize to form disulfide bonds in the course of the reaction. RNR enzymes are recycled by reaction with thioredoxin or glutaredoxin.

The regulation of RNR and related enzymes maintains a balance of dNTPs relative to each other and relative to NTPs in the cell. Very low dNTP concentration inhibits DNA synthesis and DNA repair and is lethal to the cell, while an abnormal ratio of dNTPs is mutagenic due to the increased likelihood of the DNA polymerase incorporating the wrong dNTP during DNA synthesis.

Regulation of or differential specificity of RNR has been proposed as a mechanism for alterations in the relative sizes of intracellular dNTP pools under cellular stress such as hypoxia.

In the synthesis of the nucleic acid RNA, adenosine derived from ATP is one of the four nucleotides incorporated directly into RNA molecules by RNA polymerases. The energy driving this polymerization comes from cleaving off a pyrophosphate (two phosphate groups). The process is similar in DNA biosynthesis, except that ATP is reduced to the deoxyribonucleotide dATP, before incorporation into DNA.

Amino Acid Activation in Protein Synthesis

Aminoacyl-tRNA synthetase enzymes utilise ATP as an energy source to attach a tRNA molecule to its specific amino acid, forming an aminoacyl-tRNA complex, ready for translation at ribosomes.

The energy is made available by ATP hydrolysis to adenosine monophosphate (AMP) as two phosphate groups are removed.

Binding to Proteins

Some proteins that bind ATP do so in a characteristic protein fold known as the Rossmann fold, which is a general nucleotide-binding structural domain that can also bind the coenzyme NAD. The most common ATP-binding proteins, known as kinases, share a small number of common folds; the protein kinases, the largest kinase superfamily, all share common structural features specialized for ATP binding and phosphate transfer.

ATP in complexes with proteins, in general, requires the presence of a divalent cation, almost always magnesium, which binds to the ATP phosphate groups. The presence of magnesium greatly decreases the dissociation constant of ATP from its protein binding partner without affecting the ability of the enzyme to catalyze its reaction once the ATP has bound. The presence of magnesium ions can serve as a mechanism for kinase regulation.

ATP Analogues

Biochemistry laboratories often use *in vitro* studies to explore ATP-dependent molecular processes. Enzyme inhibitors of ATP-dependent enzymes such as kinases are needed to examine the binding sites and transition states involved in ATP-dependent reactions. ATP analogs are also used in X-ray crystallography to determine a protein structure in complex with ATP, often together with other substrates. Most useful ATP analogs cannot be hydrolyzed as ATP would be; instead they trap the enzyme in a structure closely related to the ATP-bound state. Adenosine 5'-(gamma-thiotriphosphate) is an extremely common ATP analog in which one of the gamma-phosphate oxygens is replaced by a sulphur atom; this molecule is hydrolyzed at a dramatically slower rate than ATP itself and functions as an inhibitor of ATP-dependent processes. In crystallographic studies, hydrolysis transition states are modelled by the bound vanadate ion. However, caution is warranted in interpreting the results of experiments using ATP analogs, since some enzymes can hydrolyze them at appreciable rates at high concentration.

Chapter 3

Rates of Reaction

A chemical reaction is a process that leads to the transformation of one set of chemical substances to another. Classically, chemical reactions encompass changes that only involve the positions of electrons in the forming and breaking of chemical bonds between atoms, with no change to the nuclei (no change to the elements present), and can often be described by a chemical equation. Nuclear chemistry is a sub-discipline of chemistry that involves the chemical reactions of unstable and radioactive elements where both electronic and nuclear changes may occur.

The substance (or substances) initially involved in a chemical reaction are called reactants or reagents. Chemical reactions are usually characterized by a chemical change, and they yield one or more products, which usually have properties different from the reactants. Reactions often consist of a sequence of individual sub-steps, the so-called elementary reactions, and the information on the precise course of action is part of the reaction mechanism. Chemical reactions are described with chemical equations, which graphically present the starting materials, end products, and sometimes intermediate products and reaction conditions.

Chemical reactions happen at a characteristic reaction rate at a given temperature and chemical concentration, and rapid reactions are often described as spontaneous, requiring no input of extra energy other than thermal energy. Non-spontaneous reactions run so slowly that they are considered to require the input of some type of additional energy (such as extra heat, light or electricity) in order to proceed to completion (chemical equilibrium) at human time scales.

Different chemical reactions are used in combinations during chemical synthesis in order to obtain a desired product. In biochemistry, a similar series of chemical reactions form metabolic pathways. These reactions are often catalyzed by protein enzymes. These enzymes increase the rates of biochemical reactions, so that metabolic syntheses and decompositions impossible under ordinary conditions may be performed at the temperatures and concentrations present within a cell.

The general concept of a chemical reaction has been extended to reactions between entities smaller than atoms, including nuclear reactions, radioactive decays, and reactions between elementary particles as described by quantum field theory.

History

Chemical reactions such as combustion in the fire, fermentation and the reduction of ores to metals were known since antiquity. Initial theories of transformation of materials were developed by Greek philosophers, such as the Four-Element Theory of Empedocles stating that any substance is composed of the four basic elements – fire, water, air and earth. In the Middle Ages, chemical transformations were studied by Alchemists. They attempted, in particular, to convert lead into gold, for which purpose they used reactions of lead and lead-copper alloys with sulfur.

The production of chemical substances that do not normally occur in nature has long been tried, such as the synthesis of sulfuric and nitric acids attributed to the controversial alchemist Jβbir ibn Hayyβn. The process involved heating of sulfate and nitrate minerals such as copper sulfate, alum and saltpeter. In the 17th century, Johann Rudolph Glauber produced hydrochloric acid and sodium sulfate by reacting sulfuric acid and sodium chloride. With the development of the lead chamber process in 1746 and the Leblanc process, allowing large-scale production of sulfuric acid and sodium carbonate, respectively, chemical reactions became implemented into the industry. Further optimization of sulfuric acid technology resulted in the contact process in 1880s, and the Haber process was developed in 1909–1910 for ammonia synthesis.

From the 16th century, researchers including Jan Baptist van Helmont, Robert Boyle and Isaac Newton tried to establish theories of the experimentally observed chemical transformations. The phlogiston theory was proposed in 1667 by Johann Joachim Becher. It postulated the existence of a fire-like element called "phlogiston", which was contained within combustible bodies and released during combustion. This proved to be false in 1785 by Antoine Lavoisier who found the correct explanation of the combustion as reaction with oxygen from the air.

Joseph Louis Gay-Lussac recognized in 1808 that gases always react in a certain relationship with each other. Based on this idea and the atomic theory of John Dalton, Joseph Proust had developed the law of definite proportions, which later resulted in the concepts of stoichiometry and chemical equations.

Regarding the organic chemistry, it was long believed that compounds obtained from living organisms were too complex to be obtained synthetically. According to the concept of vitalism, organic matter was endowed with a "vital force" and distinguished from inorganic materials. This separation was ended however by the synthesis of urea from inorganic precursors by Friedrich Wöhler in 1828. Other chemists who brought major contributions to organic chemistry include Alexander William Williamson with his synthesis of ethers and Christopher Kelk Ingold, who, among many discoveries, established the mechanisms of substitution reactions.

Equations

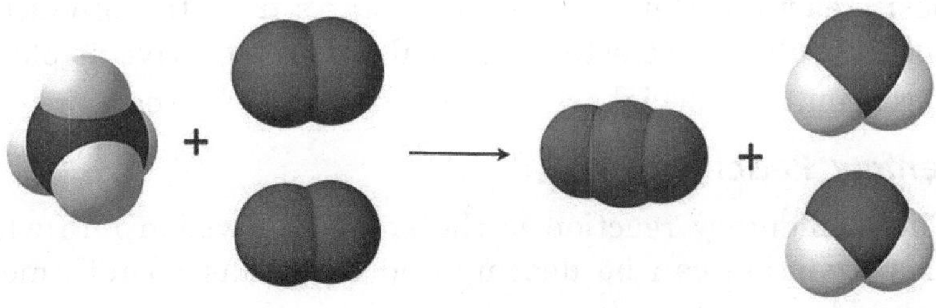

$$CH_4 + 2O_2 \longrightarrow CO_2 + 2H_2O$$

Figure: *As seen from the equation* $CH_4 + 2O_2 \rightarrow CO_2 + 2H_2O$*, a coefficient of 2 must be placed before the oxygen gas on the reactants side and before the water on the products side in order for, as per the law of conservation of mass, the quantity of each element does not change during the reaction*

Chemical equations are used to graphically illustrate chemical reactions. They consist of chemical or structural formulas of the reactants on the left and those of the products on the right. They are separated by an arrow (→) which indicates the direction and type of the reaction; the arrow is read as the word "yields". The tip of the arrow points in the direction in which the reaction proceeds. A double arrow ($\rightleftharpoons$) pointing in opposite directions is used for equilibrium reactions. Equations should be balanced according to the stoichiometry, the number of atoms of each species should be the same on both sides of the equation. This is achieved by scaling the number of involved molecules (*A, B, C* and *D* in a schematic example below) by the appropriate integers *a, b, c* and *d*.

$$\text{a A+b B} \rightarrow \text{c C+d D}$$

More elaborate reactions are represented by reaction schemes, which in addition to starting materials and products show important intermediates or transition states. Also, some relatively minor additions to the reaction can be indicated above the reaction arrow; examples of such additions are water, heat, illumination, a catalyst, etc.

Similarly, some minor products can be placed below the arrow, often with a minus sign.

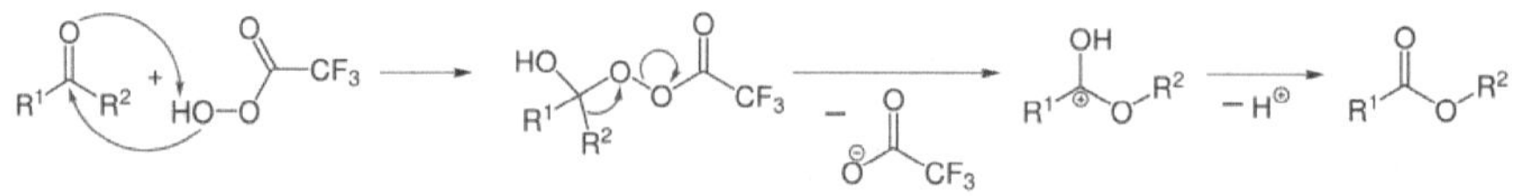

Figure: *An example of organic reaction: oxidation of ketones to esters with a peroxycarboxylic acid*

Retrosynthetic analysis can be applied to design a complex synthesis reaction. Here the analysis starts from the products, for example by splitting selected chemical bonds, to arrive at plausible initial reagents. A special arrow ($\Rightarrow$) is used in retro reactions.

Elementary Reactions

The elementary reaction is the smallest division into which a chemical reaction can be decomposed to, it has no intermediate products. Most experimentally observed reactions are built up from many elementary reactions that occur in parallel or sequentially. The actual sequence of the individual elementary reactions is known as reaction mechanism. An elementary reaction involves a few molecules, usually one or two, because of the low probability for several molecules to meet at a certain time.

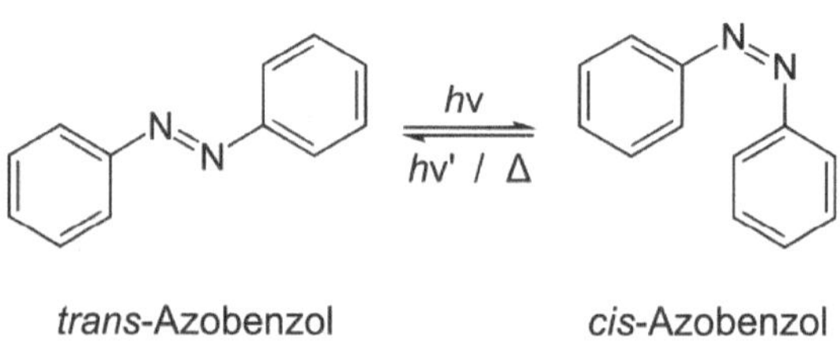

Figure: *Isomerization of azobenzene, induced by light (hv) or heat* (Δ)

The most important elementary reactions are unimolecular and bimolecular reactions. Only one molecule is involved in a unimolecular reaction; it is transformed by an isomerization or a dissociation into one or more other molecules. Such reactions require the addition of energy in the form of heat or light. A typical example of a unimolecular reaction is the cis–trans isomerization, in which the cis-form of a compound converts to the trans-form or vice versa.

In a typical dissociation reaction, a bond in a molecule splits (ruptures) resulting in two molecular fragments. The splitting can be homolytic or heterolytic. In the first case, the bond is divided so that each product retains an electron and becomes a neutral radical. In the second case, both electrons of the chemical bond remain with one of the products, resulting in charged ions. Dissociation plays an important role in triggering chain reactions, such as hydrogen–oxygen or polymerization reactions.

$$\mathrm{AB} \rightarrow \mathrm{A+B}$$

Dissociation of a molecule AB into fragments A and B

For bimolecular reactions, two molecules collide and react with each other. Their merger is called chemical synthesis or an addition reaction.

$$\mathrm{A+B} \rightarrow \mathrm{AB}$$

Another possibility is that only a portion of one molecule is transferred to the other molecule. This type of reaction occurs, for example, in redox and acid-base reactions. In redox reactions, the transferred particle is an electron, whereas in acid-base reactions it is a proton. This type of reaction is also called metathesis.

$$\mathrm{HA+B} \rightarrow \mathrm{A+HB}$$

for example

$$NaCl_{(aq)} + AgNO_{3(aq)} \rightarrow NaNO_{3(aq)} + AgCl_{(s)}$$

Chemical Equilibrium

Most chemical reactions are reversible, that is they can and do run in both directions. The forward and reverse reactions are competing with each other and differ in reaction rates. These rates depend on the concentration and therefore change with time of the reaction: the reverse rate gradually increases and becomes equal to the rate of the forward reaction, establishing the so-called chemical equilibrium. The time to reach equilibrium depends on such parameters as temperature, pressure and the materials involved, and is determined by the minimum free energy. In equilibrium, the Gibbs free energy must be zero. The pressure dependence can be explained with the Le Chatelier's principle. For example, an increase in pressure due to decreasing volume causes the reaction to shift to the side with the fewer moles of gas.

The reaction yield stabilizes at equilibrium, but can be increased by removing the product from the reaction mixture or changed by increasing the temperature or pressure. A change in the concentrations of the reactants does not affect the equilibrium constant, but does affect the equilibrium position.

Thermodynamics

Chemical reactions are determined by the laws of thermodynamics. Reactions can proceed by themselves if they are exergonic, that is if they release energy. The associated free energy of the reaction is composed of two different thermodynamic quantities, enthalpy and entropy:

$$\Delta G = \Delta H - T \cdot \Delta S.$$

G: free energy, H: enthalpy, T: temperature, S: entropy, Δ : difference(change between original and product)

Reactions can be exothermic, where ΔH is negative and energy is released. Typical examples of exothermic reactions are precipitation and crystallization, in which ordered solids are formed from disordered gaseous or liquid phases. In contrast, in endothermic reactions, heat is consumed from the environment. This can occur by increasing the entropy of the system, often through the formation of gaseous reaction products, which have high entropy. Since the entropy increases with temperature, many endothermic reactions preferably take place at high temperatures. On the contrary, many exothermic reactions such as crystallization occur at low temperatures. Changes in temperature can sometimes reverse the sign of the enthalpy of a reaction, as for the carbon monoxide reduction of molybdenum dioxide:

$$2CO_{(g)} + MoO_{2(s)} \rightarrow 2CO_{2(g)} + Mo_{(s)}; \Delta H° = +21.86 \text{ kJ at } 298 \text{ K}$$

This reaction to form carbon dioxide and molybdenum is endothermic at low temperatures, becoming less so with increasing temperature. $\Delta H°$ is zero at 1,855 K, and the reaction becomes exothermic above that temperature.

Changes in temperature can also reverse the direction tendency of a reaction. For example, the water gas shift reaction

$$CO_{(g)} + H_2O_{(v)} \rightleftharpoons CO_{2(g)} + H_{2(g)}$$

is favoured by low temperatures, but its reverse is favoured by high temperature. The shift in reaction direction tendency occurs at 1,100 K.

Reactions can also be characterized by the internal energy which takes into account changes in the entropy, volume and chemical potential. The latter depends, among other things, on the activities of the involved substances.

$$dU = TdS - pdV + \mu dn$$

U: internal energy, S: entropy, p: pressure, μ: chemical potential, n: number of molecules, d: small change sign

Kinetics

The speed at which a reactions takes place is studied by reaction kinetics. The rate depends on various parameters, such as:

- Reactant concentrations, which usually make the reaction happen at a faster rate if raised through increased collisions per unit time. Some reactions, however, have rates that are *independent* of reactant concentrations. These are called zero order reactions.
- Surface area available for contact between the reactants, in particular solid ones in heterogeneous systems. Larger surface areas lead to higher reaction rates.
- Pressure – increasing the pressure decreases the volume between molecules and therefore increases the frequency of collisions between the molecules.
- Activation energy, which is defined as the amount of energy required to make the reaction start and carry on spontaneously. Higher activation energy implies that the reactants need more energy to start than a reaction with a lower activation energy.
- Temperature, which hastens reactions if raised, since higher temperature increases the energy of the molecules, creating more collisions per unit time,
- The presence or absence of a catalyst. Catalysts are substances which change the pathway (mechanism) of a reaction which in turn increases the speed of a reaction by lowering the activation energy needed for the reaction to take place. A catalyst is not destroyed or changed during a reaction, so it can be used again.
- For some reactions, the presence of electromagnetic radiation, most notably ultraviolet light, is needed to promote the breaking of bonds to start the reaction. This is particularly true for reactions involving radicals.

Several theories allow calculating the reaction rates at the molecular level. This field is referred to as reaction dynamics. The rate v of a first-order reaction, which could be disintegration of a substance A, is given by:

$$v = -\frac{d[\mathrm{A}]}{dt} = k \cdot [\mathrm{A}].$$

Its integration yields:

$$[\mathrm{A}](t) = [\mathrm{A}]_0 \cdot e^{-k \cdot t}.$$

Here k is first-order rate constant having dimension 1/time, [A](t) is concentration at a time t and $[\mathrm{A}]_0$ is the initial concentration. The rate of a first-order reaction depends only on the concentration and the properties of the involved substance, and the reaction itself can be described with the characteristic half-life. More than one time constant is needed when describing reactions of higher order. The temperature dependence of the rate constant usually follows the Arrhenius equation:

$$k = k_0 e^{-E_a / k_B T}$$

where E_a is the activation energy and k_B is the Boltzmann constant. One of the simplest models of reaction rate is the collision theory. More realistic models are tailored to a specific problem and include the transition state theory, the calculation of the potential energy surface, the Marcus theory and the Rice–Ramsperger–Kassel–Marcus (RRKM) theory.

Reaction Types

Four Basic Types

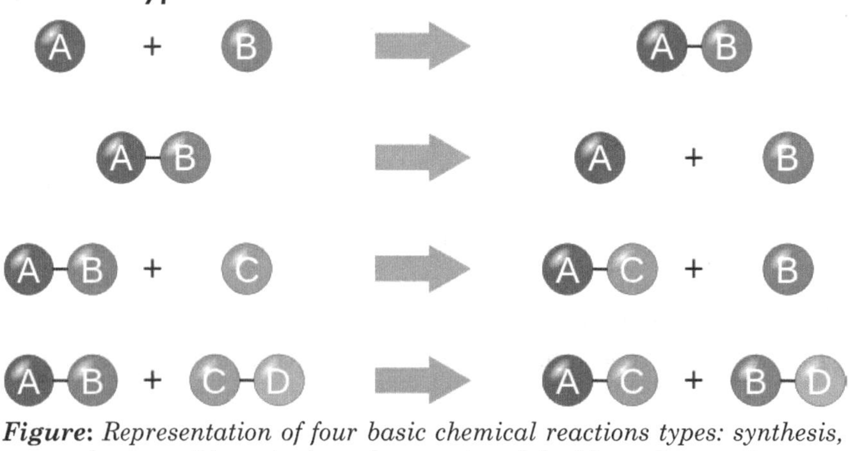

Figure: *Representation of four basic chemical reactions types: synthesis, decomposition, single replacement and double replacement.*

Synthesis

In a synthesis reaction, two or more simple substances combine to form a more complex substance. These reactions are in the general form:

$$A + B \rightarrow AB$$

Two or more reactants yielding one product is another way to identify a synthesis reaction. One example of a synthesis reaction is the combination of iron and sulfur to form iron(II) sulfide:

$$8Fe + S_8 \rightarrow 8FeS$$

Another example is simple hydrogen gas combined with simple oxygen gas to produce a more complex substance, such as water.

Decomposition

A decomposition reaction is the opposite of a synthesis reaction, where a more complex substance breaks down into its more simple parts. These reactions are in the general form:

$$AB \rightarrow A + B$$

One example of a decomposition reaction is the electrolysis of water to make oxygen and hydrogen gas:

$$2H_2O \rightarrow 2H_2 + O_2$$

Single Replacement

In a single replacement reaction, a single uncombined element replaces another in a compound; in order words, one element trades places with another element in a compound These reactions come in the general form of:

$$A + BC \rightarrow AC + B$$

One example of a single displacement reaction is when magnesium replaces hydrogen in water to make magnesium hydroxide and hydrogen gas:

$$Mg + 2H_2O \rightarrow Mg(OH)_2 + H_2$$

Double Replacement

In a double replacement reaction, the anions and cations of two compounds switch places and form two entirely different compounds. These reactions are in the general form:

$$AB + CD \rightarrow AD + CB$$

For example, when barium chloride ($BaCl_2$) and magnesium sulfate ($MgSO_4$) react, the SO_4^{2-} anion switches places with the $2Cl^-$ anion, giving the compounds $BaSO_4$ and $MgCl_2$.

Another example of a double displacement reaction is the reaction of lead(II) nitrate with potassium iodide to form lead(II) iodide and potassium nitrate:

$$Pb(NO_3)_2 + 2KI \rightarrow PbI_2 + 2KNO_3$$

Oxidation and Reduction

Redox reactions can be understood in terms of transfer of electrons from one involved species (reducing agent) to another (oxidizing agent). In this process, the former species is *oxidized* and the latter is *reduced.* Though sufficient for many purposes, these descriptions are not precisely correct. Oxidation is better defined as an increase in oxidation state, and reduction as a decrease in oxidation state. In practice, the transfer of electrons will always change the oxidation state, but there are many reactions that are classed as "redox" even though no electron transfer occurs (such as those involving covalent bonds).

In the following redox reaction, hazardous sodium metal reacts with toxic chlorine gas to form the ionic compound sodium chloride, or common table salt:

$$2Na_{(s)} + Cl_{2(g)} \rightarrow 2NaCl_{(s)}$$

In the reaction, sodium metal goes from an oxidation state of 0 (as it is a pure element) to +1: in other words, the sodium lost one electron and is said to have been oxidized. On the other hand, the chlorine gas goes from an oxidation of 0 (it is also a pure element) to -1: the chlorine gains one electron and is said to have been reduced. Because the chlorine is the one reduced, it is considered the electron acceptor, or in other words, induces oxidation in the sodium - thus the chlorine gas is considered the oxidizing agent. Conversely, the sodium is oxidized or is the electron donor, and thus induces reduction in the other species and is considered the *reducing agent.*

Which of the involved reactants would be reducing or oxidizing agent can be predicted from the electronegativity of their elements. Elements with low electronegativity, such as most metals, easily donate electrons and oxidize – they are reducing agents.

On the contrary, many ions with high oxidation numbers, such as $H_2O_2, MnO_4^-, CrO_3, Cr_2O_7^{2-}, OsO_4$ can gain one or two extra electrons and are strong oxidizing agents.

The number of electrons donated or accepted in a redox reaction can be predicted from the electron configuration of the reactant element. Elements try to reach the low-energy noble gas configuration, and therefore alkali metals and halogens will donate and accept one electron respectively. Noble gases themselves are chemically inactive.

An important class of redox reactions are the electrochemical reactions, where electrons from the power supply are used as the reducing agent. These reactions are particularly important for the production of chemical elements, such as chlorine or aluminium. The reverse process in which electrons are released in redox reactions and can be used as electrical energy is possible and used in batteries.

Complexation

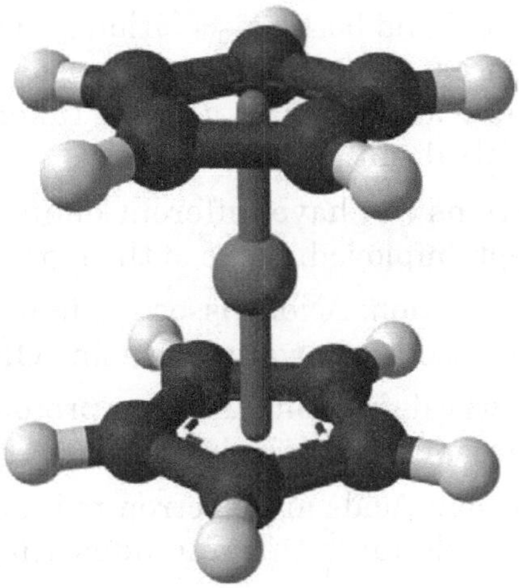

Figure: *Ferrocene – an iron atom sandwiched between two C_5H_5 ligands*

In complexation reactions, several ligands react with a metal atom to form a coordination complex. This is achieved by providing lone pairs of the ligand into empty orbitals of the metal atom and forming dipolar bonds. The ligands are Lewis bases, they can be both ions and neutral molecules, such as carbon monoxide, ammonia or water. The number of ligands that react with a central metal atom can be found using the 18-electron rule, saying that the valence shells of a transition metal will collectively accommodate 18 electrons, whereas the symmetry of the resulting complex can be predicted with the crystal field theory and ligand field theory. Complexation reactions also include ligand exchange, in which one or more ligands are replaced by another, and redox processes which change the oxidation state of the central metal atom.

Acid-Base Reactions

In the Brønsted–Lowry acid–base theory, an acid-base reaction involves a transfer of protons (H^+) from one species (the acid) to another (the base). When a proton is removed from an acid, the resulting species is termed that acid's conjugate base. When the proton is accepted by a base, the resulting species is termed that base's conjugate acid. In other words, acids act as proton donors and bases act as proton acceptors according to the following equation:

$$HA + B \rightleftharpoons A^{-} + HB^{+}$$

HA: acid, B: Base, A^-: conjugated base, HB^+: conjugated acid

The reverse reaction is possible, and thus the acid/base and conjugated base/acid are always in equilibrium. The equilibrium is determined by the acid and base dissociation constants (K_a and K_b) of the involved substances. A special case of the acid-base reaction is the neutralization where an acid and a base, taken at exactly same amounts, form a neutral salt.

Acid-base reactions can have different definitions depending on the acid-base concept employed. Some of the most common are:

- Arrhenius definition: Acids dissociate in water releasing H_3O^+ ions; bases dissociate in water releasing OH^- ions.
- Brønsted-Lowry definition: Acids are proton (H^+) donors, bases are proton acceptors; this includes the Arrhenius definition.
- Lewis definition: Acids are electron-pair acceptors, bases are electron-pair donors; this includes the Brønsted-Lowry definition.

Precipitation

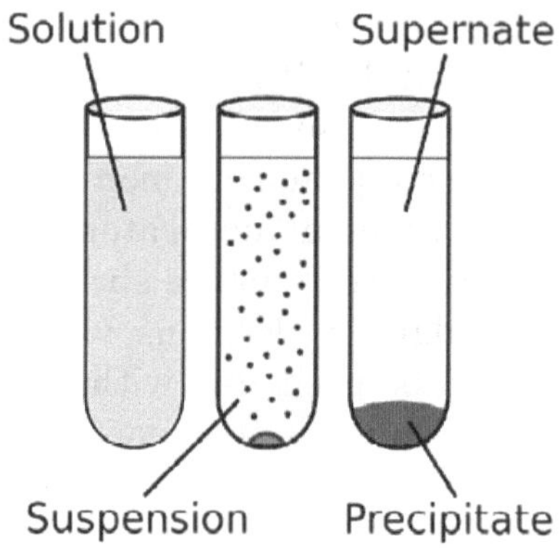

Figure: *Precipitation*

Precipitation is the formation of a solid in a solution or inside another solid during a chemical reaction. It usually takes place when the concentration of dissolved ions exceeds the solubility limit and forms an insoluble salt. This process can be assisted by adding a precipitating agent or by removal of the solvent. Rapid precipitation results in an amorphous or microcrystalline residue and slow process can yield single crystals. The latter can also be obtained by recrystallization from microcrystalline salts.

Solid-State Reactions

Reactions can take place between two solids. However, because of the relatively small diffusion rates in solids, the corresponding chemical reactions are very slow in comparison to liquid and gas phase reactions. They are accelerated by increasing the reaction temperature and finely dividing the reactant to increase the contacting surface area.

Reactions at the Solid | Gas Interface

Reaction can take place at the solid | gas interface, surfaces at very low pressure such as ultra-high vacuum. Via scanning tunneling microscopy, it is possible to observe reactions at the solid | gas interface in real space, if the time scale of the reaction is in the correct range. Reactions at the solid | gas interface are in some cases related to catalysis.

Photochemical Reactions

In photochemical reactions, atoms and molecules absorb energy (photons) of the illumination light and convert into an excited state. They can then release this energy by breaking chemical bonds, thereby producing radicals. Photochemical reactions include hydrogen–oxygen reactions, radical polymerization, chain reactions and rearrangement reactions.

Many important processes involve photochemistry. The premier example is photosynthesis, in which most plants use solar energy to convert carbon dioxide and water into glucose, disposing of oxygen as a side-product. Humans rely on photochemistry for the formation of vitamin D, and vision is initiated by a photochemical reaction of rhodopsin. In fireflies, an enzyme in the abdomen catalyzes a reaction that results in bioluminescence. Many significant photochemical reactions, such as ozone formation, occur in the Earth atmosphere and constitute atmospheric chemistry.

Catalysis

In catalysis, the reaction does not proceed directly, but through a third substance known as catalyst. Unlike other reagents that participate in the chemical reaction, a catalyst is not consumed by the reaction itself; however, it can be inhibited, deactivated or destroyed by secondary processes. Catalysts can be used in a different phase (heterogeneous) or in the same phase (homogeneous) as the reactants. In heterogeneous catalysis, typical secondary processes include coking where the catalyst becomes covered by polymeric side products. Additionally, heterogeneous catalysts can dissolve into the solution in a solid–liquid system or evaporate in a solid–gas system.

Catalysts can only speed up the reaction – chemicals that slow down the reaction are called inhibitors. Substances that increase the activity of catalysts are called promoters, and substances that deactivate catalysts are called catalytic poisons. With a catalyst, a reaction which is kinetically inhibited by a high activation energy can take place in circumvention of this activation energy.

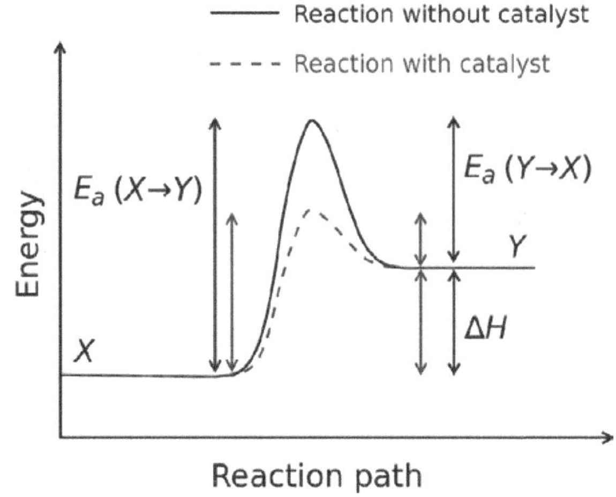

Figure: *Schematic potential energy diagram showing the effect of a catalyst in an endothermic chemical reaction. The presence of a catalyst opens a different reaction pathway (in red) with a lower activation energy. The final result and the overall thermodynamics are the same.*

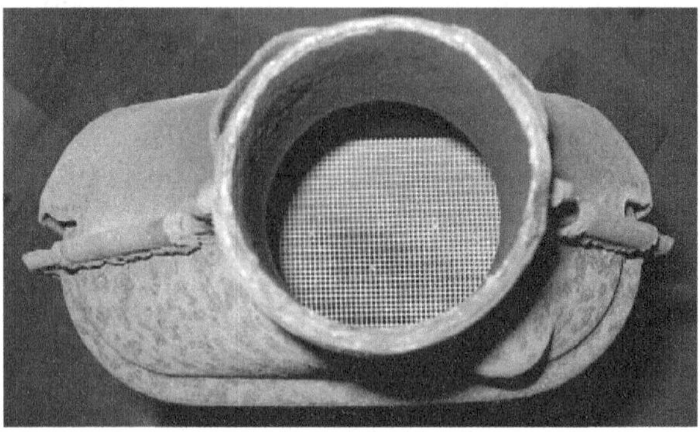

Figure: *Solid heterogeneous catalysts are plated on meshes in ceramic catalytic converters in order to maximize their surface area. This exhaust converter is from a Peugeot 106 S2 1100*

Heterogeneous catalysts are usually solids, powdered in order to maximize their surface area. Of particular importance in heterogeneous catalysis are the platinum group metals and other transition metals, which are used in hydrogenations, catalytic reforming and in the synthesis of commodity chemicals such as nitric acid and ammonia. Acids are an example of a homogeneous catalyst, they increase the nucleophilicity of carbonyls, allowing a reaction that would not otherwise proceed with electrophiles. The advantage of homogeneous catalysts is the ease of mixing them with the reactants, but they may also be difficult to separate from the products. Therefore, heterogeneous catalysts are preferred in many industrial processes.

Reactions in Organic Chemistry

In organic chemistry, in addition to oxidation, reduction or acid-base reactions, a number of other reactions can take place which involve covalent bonds between carbon atoms or carbon and heteroatoms (such as oxygen, nitrogen, halogens, etc.). Many specific reactions in organic chemistry are name reactions designated after their discoverers.

Classifications

Organic chemistry has a strong tradition of naming a specific reaction to its inventor or inventors and a long list of so-called named reactions exists, conservatively estimated at 1000. A very old named reaction is the Claisen rearrangement (1912) and a recent named reaction is the Bingel reaction (1993). When the named reaction is difficult to pronounce or very long as in the Corey-House-Posner-Whitesides reaction it helps to use the abbreviation as in the CBS reduction. The number of reactions hinting at the actual process taking place is much smaller, for example the ene reaction or aldol reaction.

Another approach to organic reactions is by type of organic reagent, many of them inorganic, required in a specific transformation. The major types are oxidizing agents such as osmium tetroxide, reducing agents such as Lithium aluminium hydride, bases such as lithium diisopropylamide and acids such as sulfuric acid.

Fundamentals

Factors governing organic reactions are essentially the same as that of any chemical reaction. Factors specific to organic reactions are those that determine the stability of reactants and products such as conjugation, hyperconjugation and aromaticity and the presence and stability of reactive intermediates such as free radicals, carbocations and carbanions.

An organic compound may consist of many isomers. Selectivity in terms of regioselectivity, diastereoselectivity and enantioselectivity is therefore an important criterion for many organic reactions. The stereochemistry of pericyclic reactions is governed by the Woodward–Hoffmann rules and that of many elimination reactions by the Zaitsev's rule.

Organic reactions are important in the production of pharmaceuticals. In a 2006 review it was estimated that 20% of chemical conversions involved alkylations on nitrogen and oxygen atoms, another 20% involved placement and removal of protective groups, 11% involved formation of new carbon-carbon bond and 10% involved functional group interconversions.

Organic Reactions by Mechanism

There is no limit to the number of possible organic reactions and mechanisms. However, certain general patterns are observed that can be used to describe many common or useful reactions. Each reaction has a stepwise reaction mechanism that explains how it happens, although this detailed description of steps is not always clear from a list of reactants alone. Organic reactions can be organized into several basic types. Some reactions fit into more than one category. For example, some substitution reactions follow an addition-elimination pathway. This overview isn't intended to include every single organic reaction. Rather, it is intended to cover the basic reactions.

Reaction type	***Subtype***	***Comment***
Addition reactions	electrophilic addition	include such reactions as halogenation, hydrohalogenation and hydration.
	nucleophilic addition	
	radical addition	
Elimination reaction		include processes such as dehydration and are found to follow an E1, E2 or E1cB reaction mechanism
Substitution reactions	nucleophilic aliphatic substitution	with S_N1, S_N2 and S_Ni reaction mechanisms
	nucleophilic aromatic substitution	
	nucleophilic acyl substitution	
	electrophilic substitution	
	electrophilic aromatic substitution	
	radical substitution	
Organic redox reactions		are redox reactions specific to organic compounds and are very common.
Rearrangement reactions	1,2-rearrangements	
	pericyclic reactions	
	metathesis	

In condensation reactions a small molecule, usually water, is split off when two reactants combine in a chemical reaction. The opposite reaction, when water is consumed in a reaction, is called hydrolysis. Many Polymerization reactions are derived from organic reactions. They are divided into addition polymerizations and step-growth polymerizations.

In general the stepwise progression of reaction mechanisms can be represented using arrow pushing techniques in which curved arrows are used to track the movement of electrons as starting materials transition to intermediates and products.

Organic Reactions by Functional Groups

Organic reactions can be categorized based on the type of functional group involved in the reaction as a reactant and the functional group that is formed as a result of this reaction. For example in the Fries rearrangement the reactant is an ester and the reaction product an alcohol.

An overview of functional groups with their preparation and reactivity is presented below:

Functional Group	*Preparation*	*Reactions*
Acid anhydride	preparation	reactions
Acyl halides	preparation	reactions
Acyloins	preparation	reactions
Alcohols	preparation	reactions
Aldehydes	preparation	reactions
Alkanes	preparation	reactions
Alkenes	preparation	reactions
Alkyl halides	preparation	reactions
Alkyl nitrites	preparation	reactions
Alkynes	preparation	reactions
Amides	preparation	reactions
Amine oxide	preparation	reactions
Amines	preparation	reactions
Arene compounds	preparation	reactions
Azides	preparation	reactions
Aziridines	preparation	reactions
Carboxylic acids	preparation	reactions
Cyclopropanes	preparation	reactions
Diazo compounds	preparation	reactions
Diols	preparation	reactions

Contd...

Functional Group	***Preparation***	***Reactions***
Esters	preparation	reactions
Ethers	preparation	reactions
Epoxide	preparation	reactions
Haloketones	preparation	reactions
Imines	preparation	reactions
Isocyanates	preparation	reactions
Ketones	preparation	reactions
Lactams	preparation	reactions
Nitriles	preparation	reactions
Nitro compounds	preparation	reactions
Phenols	preparation	reactions
Thiols	preparation	reactions

Other Organic Reaction Classification

In heterocyclic chemistry, organic reactions are classified by the type of heterocycle formed with respect to ring-size and type of heteroatom. Reactions are also categorized by the change in the carbon framework. Examples are ring expansion and ring contraction, homologation reactions, polymerization reactions, insertion reactions, ring-opening reactions and ring-closing reactions.

Organic reactions can also be classified by the type of bond to carbon with respect to the element involved. More reactions are found in organosilicon chemistry, organosulfur chemistry, organophosphorus chemistry and organofluorine chemistry. With the introduction of carbon-metal bonds the field crosses over to organometallic chemistry.

Substitution

In a substitution reaction, a functional group in a particular chemical compound is replaced by another group. These reactions can be distinguished by the type of substituting species into a nucleophilic, electrophilic or radical substitution.

In the first type, a nucleophile, an atom or molecule with an excess of electrons and thus a negative charge or partial charge, replaces another atom or part of the "substrate" molecule. The electron pair from the nucleophile attacks the substrate forming a new bond, while the leaving group departs with an electron pair. The nucleophile may be electrically neutral or negatively charged, whereas the substrate is typically neutral or positively charged. Examples of nucleophiles are hydroxide ion, alkoxides, amines and halides. This type of reaction is found mainly in aliphatic hydrocarbons, and rarely in aromatic hydrocarbon. The latter have high electron density and enter nucleophilic aromatic substitution only with very strong electron withdrawing groups. Nucleophilic substitution can take place by two different mechanisms, S_N1 and S_N2. In their names, S stands for substitution, N for nucleophilic, and the number represents the kinetic order of the reaction, unimolecular or bimolecular.

The S_N1 reaction proceeds in two steps. First, the leaving group is eliminated creating a carbocation. This is followed by a rapid reaction with the nucleophile.

In the S_N2 mechanism, the nucleophile forms a transition state with the attacked molecule, and only then the leaving group is cleaved. These two mechanisms differ in the stereochemistry of the products. S_N1 leads to the non-stereospecific addition and does not result in a chiral centre, but rather in a set of geometric isomers (*cis/trans*). In contrast, a reversal (Walden inversion) of the previously existing stereochemistry is observed in the S_N2 mechanism.

Electrophilic substitution is the counterpart of the nucleophilic substitution in that the attacking atom or molecule, an electrophile, has low electron density and thus a positive charge. Typical electrophiles are the carbon atom of carbonyl groups, carbocations or sulfur or nitronium cations.

This reaction takes place almost exclusively in aromatic hydrocarbons, where it is called electrophilic aromatic substitution. The electrophile attack results in the so-called σ-complex, a transition state in which the aromatic system is abolished. Then, the leaving group, usually a proton, is split off and the aromaticity is restored. An alternative to aromatic substitution is electrophilic aliphatic substitution. It is similar to the nucleophilic aliphatic substitution and also has two major types, S_E1 and S_E2.

In the third type of substitution reaction, radical substitution, the attacking particle is a radical. This process usually takes the form of a chain reaction, for example in the reaction of alkanes with halogens. In the first step, light or heat disintegrates the halogen-containing molecules producing the radicals. Then the reaction proceeds as an avalanche until two radicals meet and recombine.

$$X\cdot + R\text{-}H \rightarrow X\text{-}H + R\cdot$$

$$R\cdot + X_2 \rightarrow R\text{-}X + X\cdot$$

Reactions during the chain reaction of radical substitution

Addition and Elimination

The addition and its counterpart, the elimination, are reactions which change the number of substitutents on the carbon atom, and form or cleave multiple bonds. Double and triple bonds can be produced by eliminating a suitable leaving group. Similar to the nucleophilic substitution, there are several possible reaction mechanisms which are named after the respective reaction order. In the E1 mechanism, the leaving group is ejected first, forming a carbocation. The next step, formation of the double bond, takes place with elimination of a proton (deprotonation). The leaving order is reversed in the E1cb mechanism, that is the proton is split off first. This mechanism requires participation of a base. Because of the similar conditions, both reactions in the E1 or E1cb elimination always compete with the S_N1 substitution.

The E2 mechanism also requires a base, but there the attack of the base and the elimination of the leaving group proceed simultaneously and produce no ionic intermediate. In contrast to the E1 eliminations, different stereochemical configurations are possible for the reaction product in the E2 mechanism, because the attack of the base preferentially occurs in the anti-position with respect to the leaving group. Because of the similar conditions and reagents, the E2 elimination is always in competition with the S_N2-substitution.

The counterpart of elimination is the addition where double or triple bonds are converted into single bonds. Similar to the substitution reactions, there are several types of additions distinguished by the type of the attacking particle. For example, in the electrophilic addition of hydrogen bromide, an electrophile (proton) attacks the double bond forming a carbocation, which then reacts with the nucleophile (bromine). The carbocation can be formed on either side of the double bond depending on the groups attached to its ends, and the preferred configuration can be predicted with the Markovnikov's rule. This rule states that "In the heterolytic addition of a polar molecule to an alkene or alkyne, the more electronegative (nucleophilic) atom (or part) of the polar molecule becomes attached to the carbon atom bearing the smaller number of hydrogen atoms."

If the addition of a functional group takes place at the less substituted carbon atom of the double bond, then the electrophilic substitution with acids is not possible. In this case, one has to use the hydroboration–oxidation reaction, where in the first step, the boron atom acts as electrophile and adds to the less substituted carbon atom. At the second step, the nucleophilic hydroperoxide or halogen anion attacks the boron atom.

While the addition to the electron-rich alkenes and alkynes is mainly electrophilic, the nucleophilic addition plays an important role for the carbon-heteroatom multiple bonds, and especially its most important representative, the carbonyl group. This process is often associated with an elimination, so that after the reaction the carbonyl group is present again. It is therefore called addition-elimination reaction and may occur in carboxylic acid derivatives such as chlorides, esters or anhydrides. This reaction is often catalyzed by acids or bases, where the acids increase by the electrophilicity of the carbonyl group by binding to the oxygen atom, whereas the bases enhance the nucleophilicity of the attacking nucleophile.

Nucleophilic addition of a carbanion or another nucleophile to the double bond of an alpha, beta unsaturated carbonyl compound can proceed via the Michael reaction, which belongs to the larger class of conjugate additions. This is one of the most useful methods for the mild formation of C–C bonds.

Some additions which can not be executed with nucleophiles and electrophiles, can be succeeded with free radicals. As with the free-radical substitution, the radical addition proceeds as a chain reaction, and such reactions are the basis of the free-radical polymerization.

Other Organic Reaction Mechanisms

In a rearrangement reaction, the carbon skeleton of a molecule is rearranged to give a structural isomer of the original molecule. These include hydride shift reactions such as the Wagner-Meerwein rearrangement, where a hydrogen, alkyl or aryl group migrates from one carbon to a neighbouring carbon. Most rearrangements are associated with the breaking and formation of new carbon-carbon bonds. Other examples are sigmatropic reaction such as the Cope rearrangement.

Cyclic rearrangements include cycloadditions and, more generally, pericyclic reactions, wherein two or more double bond-containing molecules form a cyclic molecule. An important example of cycloaddition reaction is the Diels–Alder reaction (the so-called [4+2] cycloaddition) between a conjugated diene and a substituted alkene to form a substituted cyclohexene system.

Whether or not a certain cycloaddition would proceed depends on the electronic orbitals of the participating species, as only orbitals with the same sign of wave function will overlap and interact constructively to form new bonds. Cycloaddition is usually assisted by light or heat. These perturbations result in different arrangement of electrons in the excited state of the involved molecules and therefore in different effects. For example, the [4+2] Diels-Alder reactions can be assisted by heat whereas the [2+2] cycloaddition is selectively induced by light. Because of the orbital character, the potential for developing stereoisomeric products upon cycloaddition is limited, as described by the Woodward–Hoffmann rules.

Biochemical Reactions

Biochemical reactions are mainly controlled by enzymes. These proteins can specifically catalyze a single reaction, so that reactions can be controlled very precisely. The reaction takes place in the active site, a small part of the enzyme which is usually found in a cleft or pocket lined by amino acid residues, and the rest of the enzyme is used mainly for stabilization. The catalytic action of enzymes relies on several mechanisms including the molecular shape ("induced fit"), bond strain, proximity and orientation of molecules relative to the enzyme, proton donation or withdrawal (acid/base catalysis), electrostatic interactions and many others.

The biochemical reactions that occur in living organisms are collectively known as metabolism. Among the most important of its mechanisms is the anabolism, in which different DNA and enzyme-controlled processes result in the production of large molecules such as proteins and carbohydrates from smaller units. Bioenergetics studies the sources of energy for such reactions. An important energy source is glucose, which can be produced by plants via photosynthesis or assimilated from food. All organisms use this energy to produce adenosine triphosphate (ATP), which can then be used to energize other reactions.

Applications

Chemical reactions are central to chemical engineering where they are used for the synthesis of new compounds from natural raw materials such as petroleum and mineral ores. It is essential to make the reaction as efficient as possible, maximizing the yield and minimizing the amount of reagents, energy inputs and waste. Catalysts are especially helpful for reducing the energy required for the reaction and increasing its reaction rate.

Some specific reactions have their niche applications. For example, the thermite reaction is used to generate light and heat in pyrotechnics and welding. Although it is less controllable than the more conventional oxy-fuel welding, arc welding and flash welding, it requires much less equipment and is still used to mend rails, especially in remote areas.

Monitoring

Mechanisms of monitoring chemical reactions depend strongly on the reaction rate. Relatively slow processes can be analyzed in situ for the concentrations and identities of the individual ingredients. Important tools of real time analysis are the measurement of pH and analysis of optical absorption (colour) and emission spectra.

A less accessible but rather efficient method is introduction of a radioactive isotope into the reaction and monitoring how it changes over time and where it moves to; this method is often used to analyze redistribution of substances in the human body. Faster reactions are usually studied with ultrafast laser spectroscopy where utilization of femtosecond lasers allows short-lived transition states to be monitored at time scaled down to a few femtoseconds.

Transport Processes

Organisms need to be able to move materials (such as respiratory gases, nutrients, waste products and heat) both into and out of, and within, themselves - that is called transport.

Organisms do not always require transport systems, and most have much simpler ones than us - so why the complexity in the mammalian transport system? The reason is our size. Because mammals are so large (increased distance from the nutrients and the cells requiring them), have a high metabolic rate and a high level of activity, we have high oxygen and nutrient requirements. We also produce a relatively large amount of waste that has to be removed - all this is achieved by our complex transport system with pump.

Table 1: Transport systems

Type of organism	*Single-celled*	*Jellyfish and sea anemones*	*Insects*	*Green plants*	*Fish*	*Mammals*
Size Range	Microscopic	Up to 60cm	1mm to 13cm	1mm to 150m	12mm to 10m	35mm to 34m
Example	E.Coli	Sea anemone	Locust	Oak tree	Minnow	Human
Activity Level	Move using flagellum in search of food	Anemones are sedantry, jellyfish swim slowly	Active movement, many fly	No movement	Active movement	Most active movement
Transport system	No specialised transport system	No specialised transport system	Blood system (has pumps)	Xylem/phloem (no pump)	Blood system (has pump)	Blood system (has pump)

As you can see from the above table, some organisms do not have transport systems, and rely on diffusion alone since they are such simple creatures that diffusion is adequate. Relative to their volume, they have a large surface area for diffusion to occur. Jellyfish and sea anemones survive for the same two reasons and that they do not move around much (which requires oxygen).

Cardiovascular System

The cardiovascular system or simply the blood system has a pump and a series of tubes, the heart and blood vessels respectively. The blood, coming from the left ventricle of the heart travels twice through the heart to make a complete circuit. Starting at the left ventricle, into the aorta, all around the body except for the lungs, back into the right side of the heart through the vena cava and then pumped out of the right ventricle into the pulmonary artery, carrying it to the lungs to be reoxygenated, then returning along the pulmonary vein to the left side of the heart. This combination makes it a double circulatory system, and since the blood never leaves it is known as a closed circulatory system.

Arteries

The function of the arteries is to transport blood swiftly and at high blood pressure to the tissues. Look at the diagram on the right. It is of a artery and shows three things - the tunica intima, the tunica media and the tunica externa. The tunica intima is made from endothelium, resting on a thin layer of elastic fibres as this is in contact with the blood this is vital. Next up is the middle layer called the tunica media, containing smooth muscle, collagen and elastic fibres. Finally, the outside, the tunica externa, containing elastic and collagen fibres. The space in the middle where the blood flows through is called the lumen.

Arteries have the thickest walls of any blood vessel, this, along with the smooth muscle and elastic fibres allow the wall to stretch as pulses of blood from the heart come through at high pressure. The fact that they are elastic (i.e., that they recoil after being stretched) is vital to ensure they 'snap back' after the pulse of blood has been through, raising the pressure and giving the blood a push.

As they get further from the heart, and closer to the tissue to which they are giving blood to, they branch into smaller vessels known as arterioles - similar to arteries but smaller and with more muscle to push the blood along if required, also to enhance or restrict blood flow, such as restricting blood flow to gut during exercise.

Veins

You may be asked to look at a diagram drawn as if you were looking into the vein, such as Veins have a relatively large lumen, a thin tunica intima and an even thinner tunica media. Their tunica externa is mostly collagen fibres, as opposed to the artery that has a lot more elasticised fibres.

Blood flows from the arterioles into capillaries, and when it leaves there it enters the veins, whose function it is to return blood to the heart. Veins have to deal with very low pressure blood, typically less than 5 mm hg - this helpfully negates the need for thick walls but how can blood be returned to the heart under such low pressure?

The answer is semi-lunar valves. As you can see from the picture, the semi-lunar valve only allows blood to go one way, trying to go the other simply closes the valve. Muscles in your legs also help to raise the pressure within your veins.

Capillaries

The arterioles continue to branch from the artery, eventually forming the smallest of all blood vessels, capillaries. The capillaries function is to take blood as close as possible to cells, allowing rapid transport of substances between cells and blood - they form a network in every tissue known as a capillary.

Capillary walls are extremely thin, with walls of just a single layer of endothelial cells and each capillary is about the same size as a red blood cell, 7μm. This structure allows blood get to as close as 1μm from individual cells. Most capillary beds will have gaps in between the individual cells to allow some components of the blood to seep into the space between the cells of the body, forming tissue fluid. The blood pressure in a capillary is around 10mm Hg.

Blood and Tissue Fluid

Blood is simply cells floating in a pale yellow liquid called blood plasma. Blood plasma is mostly water with solutes such as nutrients and waste products. Protein molecules that remain in the blood all the time are known as plasma proteins. As said in the previous section, plasma leaks out between the capillaries and seeps into the spaces between the cells of the tissues, and becomes known as tissue fluid.

Tissue fluid is blood plasma, without a few things that cannot fit through capillary walls- red blood cells and protein molecules. It forms the immediate environment of each individual body cell, and exchange the materials between cells - many processes in our body are designed to maintain the composition of tissue fluid, to ensure cells have an optimum environment. This is known as homeostasis, and involves the regulation of glucose concentration, water, pH, waste products and temperature.

Lymphatic System

Of the tissue fluid discussed in the previous section, around 90% seeps back into the capillaries, and the remaining 10% becomes lymph fluid in the lymph vessels (lymphatics).

These vessels are tiny 'blind-ending' vessels, and tissue fluid can flow into them but not out - and these valves are wide enough to allow large protein molecules to pass through - important since this protein cannot get into the capillaries and can't be taken away by the blood, which would cause a fatal build-up if the tissue fluid does not take it away.

The lymph vessels join up to form larger lymph vessels, which gradually transport the lymph fluid back to the large veins which run beneath the collarbone, known as the subclavian veins.

At intervals along lymph vessels, lymph nodes lie - bacteria and other unwanted particles are removed from lymph by some white blood cells as the lymph passes through the nodes, and white blood cells within the nodes secrete.

Blood

Blood is a specialized bodily fluid composed of blood cells suspended in a liquid called blood plasma.

Blood Cells

A blood cell (also called blood corpuscle) is any cell of any type normally found in blood. In mammals, these fall into two general categories:

- Red blood cells
- White blood cells

Red

Red blood cells (erythrocytes) are red because of the pigment haemoglobin, a globular protein, the main function of which is to transport oxygen from lungs to respiring tissues. Red blood cells are relatively short-lived, they become more fragile as time goes on and rupture within the circulatory system but are replaced by the bone marrow. A list of their structures and functions are below.

'Red Blood Cells

Structure	*Function*
Small size (7mm)	To allow the red blood cell to get near to the cells to exchange oxygen with them - to fit into capillaries.
No nucleus, mitochondria or endoplasmic reticulum.	Red blood cells are highly specialized cells, and they provide as much space for haemoglobin as possible.
Biconcave disk shape	Large surface area for oxygen diffusion. Flexibility.

White

White blood cells (leucocytes) are also made in the bone marrow, but differ from red blood cells in the following ways.

- They always have a nucleus
- Larger than red blood
- Not bi-concave, usually spherical or just irregular.

There are two main groups of white blood cells - phagocytes and lymphocytes.

Phagocytes

- Destroy invading cells by phagocytosis
- Recognised by their lobed nuclei and granular cytoplasm.

Lymphocytes

- Destroy microorganisms but not by phagocytosis
- Secrete antibodies
- Smaller than phagocytes
- Large nucleus
- Small cytoplasm

Haemoglobin

The main role of the transport system is to transport oxygen from the alveoli to where it is needed - cells around the body, and this is achieved in the protein haemoglobin. Each haemoglobin molecule can hold four oxygen molecules.

Dissociation Curve

Haemoglobin must not only pickup oxygen at the lungs (as we have seen so far) but to drop it off at the appropriate tissues. This is known as dissociation, and it is possible to measure haemoglobins dissociation and produce a graph, known as a dissociation curve.

The graph (seen here on the right), shows that at low partial pressures of oxygen, the oxygen-saturation of haemoglobin is very low. At high partial pressures of oxygen, the oxygen-saturation of haemoglobin is very high.

S-Shaped

The S-Curve can be explained by the behaviour of a haemoglobin molecule as it loses and gains oxygen molecules.

A haemoglobin molecule has four haem groups, and when an oxygen molecule combines with a haem group, the haemoglobin molecule becomes slightly distorted, and this distortion increases the haemoglobins affinity for oxygen, that is, it makes it easier for the second molecule of oxygen to bind. The second oxygen molecule joins, in turn making it easier for a third oxygen molecule to bind, but this molecule makes it harder for the fourth to bind.

All this explains why the initial rise is slow, (first is hard to bind but the 25th percentile to the 75th percentile (2nd and 3rd are easy) is a steep graph, and the last quarter of the graph is slow (fourth is hard) again.

The Bohr Effect

The Bohr effect is the influence of carbon dioxide on the S-shaped graph - when carbon dioxide is released into the red blood cells, it is converted by the enzyme carbonic anhydrase, which produces excess hydrogen ions as a result. Haemoglobin readily combines with these ions, forming haemoglobinic acid, and in doing so releases the oxygen which it is carrying.

This results in two things - haemoglobin mopping up the hydrogen ions which are formed when carbon dioxide dissolves and dissociates - keeping the pH (concentration of hydrogen ions) neutral (acting as a buffer).

The fact that a high partial pressure of carbon dioxide causes haemoglobin to release oxygen - this is the Bohr effect. It is very useful in that high concentrations of carbon dioxide are found in actively respiring tissues - the ones that need oxygen the most, and cause haemoglobin to release oxygen even more readily that it would otherwise do. The Bohr Effect shifts the S curve (this time for dissociation) slightly to the right, simply meaning that haemoglobin is less saturated than it would be at a low partial pressure of carbon dioxide.

CO_2 Transport

There are three ways CO_2 can be transported in the blood.

- Carbon dioxide can be carried directly in solution. (5%)
- One of the products of carbon dioxide dissociation is hydrogencarbonate ions, which mostly diffuse into the blood plasma. (85%)
- Carbon dioxide can also combine directly with haemoglobin to form carbaminohaemoglobin. (10%)

When the blood reaches the lungs, the above first two reactions are reversed and this leaves carbon dioxide to diffuse out of the blood, and the haemoglobin free to combine with oxygen.

Foetal Haemoglobin

A developing foetus receives its oxygen across the placenta from its mother's haemoglobin. Obviously, the mother's blood has to supply her whole body as well as the foetus' needs, and thus by the time the blood reaches the placenta, the partial pressure of oxygen is relatively low.

This requires foetal haemoglobin to have a higher affinity for oxygen, and bind more readily at lower partial pressures of oxygen. The curve for foetal haemoglobin on the s-shaped curve graph is slightly to the left of the adult haemoglobin curve.

Once the foetus is born, he/she loses their foetal haemoglobin within 6 months, so that in the future a female can have a foetus inside her and that foetus will be able to use its foetal haemoglobin to attain oxygen.

Also, it is necessary that the foetal haemoglobin changes to adult haemoglobin, so that the oxygen affinity is lowered sufficiently enough for the right amount of oxygen to be given up to cells and tissues. This is especially important as the child becomes more active, because its tissues will require more oxygen.

Myoglobin

Myoglobin is a reddish pigment which combines with oxygen, just like haemoglobin. However it is mostly found in muscle tissue. It has only one polypeptide chain, one haem group and can only bind with one oxygen molecule.

However, myoglobin has a very high oxygen affinity and will not release its oxygen unless the partial pressure of oxygen around it is very low.

This is useful because during the initial minutes of exercise, the heart and lungs require time to catch up with the muscles demand, and during this time the oxygen saturation drops low in the muscles as they quickly use it, and myoglobin releases its oxygen. So, myoglobin acts as an oxygen store.

Oxygen Transport Difficulties

As you probably have deduced, oxygen transport in our bodies is incredibly efficient, but it can be affected by a few things.

Carbon Monoxide

Haemoglobin, for all its efficiency has a flaw—it combines, irreversibly, with carbon monoxide with an affinity 250 times that of oxygen. Carbon monoxide is inhaled from fumes from many sources, and combines with haemoglobin to form carboxyhaemoglobin.

Thus excessive concentrations of carbon monoxide, like from poorly-ventilated gas heaters, severely impact the bodies' oxygen carrying capacity, and carbon monoxide poisoning can lead to death from asphyxiation.

Altitude Sickness

Since haemoglobin partially relies atmospheric pressure to bind readily with oxygen, humans sometimes encounter problems above heights of 6500 feet, as the air pressure becomes such that haemoglobin will at most become 70% saturated instead of the usual 92-95% saturated, causing less oxygen to be carried around the body.

This can make people feel ill, but worse it causes the arterioles in their brain to dilate, and increase the amount of blood flowing into capillaries. This causes fluid to leak from the capillaries into the brain tissues causing disorientation, and can even leak to the lungs making it difficult to breathe. This condition can be fatal.

However, given time, the body can acclimatise to the lower pressure by increasing the number of red blood cells—however this takes at least two or three weeks at a high altitude. Other changes that occur to people who live at high altitudes include broad chests (high lung capacity), larger hearts and more haemoglobin in the blood than usual.

Transport in Plants

Plants don't have a circulatory system like animals, but they do have a sophisticated transport system for carrying water and dissolved solutes to different parts of the plant, often over large distances.

Stem Structure

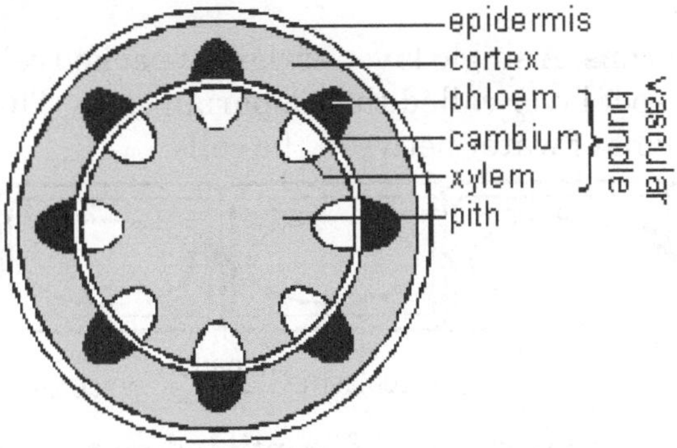

- Epidermis. One cell thick. In young plants the epidermis cells may secrete a waterproof cuticle, and in older plants the epidermis may be absent, replaced by bark.
- Cortex. Composed of various "packing" cells, to give young plants strength and flexibility, and are the source of plant fibres such as sisal and hemp.
- Vascular Tissue. This contains the phloem and xylem tissue, which grow out from the cambium. In dicot plants (the broad-leafed plants), the vascular tissue is arranged in vascular bundles, with phloem on the outside and xylem on the inside. In older plants the xylem bundles fuse together to form the bulk of the stem.
- Pith. The central region of a stem, used for food storage in young plants. It may be absent in older plants (i.e. they're hollow).

Root Structure

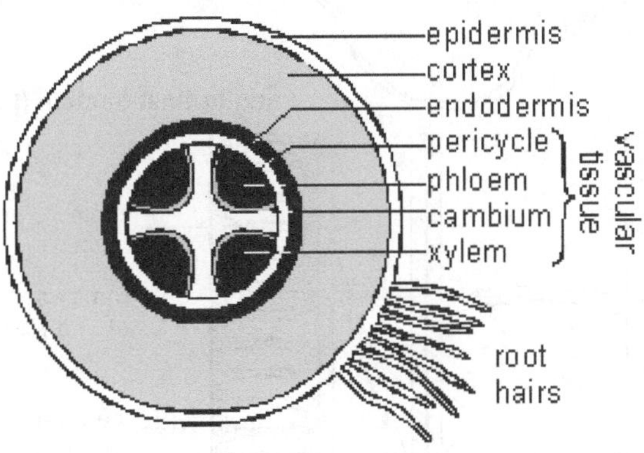

- Epidermis. A single layer of cells often with long extensions called root hairs, which increase the surface area enormously. A single plant may have 10^{10} root hairs.

- Cortex. A thick layer of packing cells often containing stored starch.
- Endodermis. A single layer of tightly-packed cells containing a waterproof layer called the casparian strip. This prevents the movement of water between the cells.

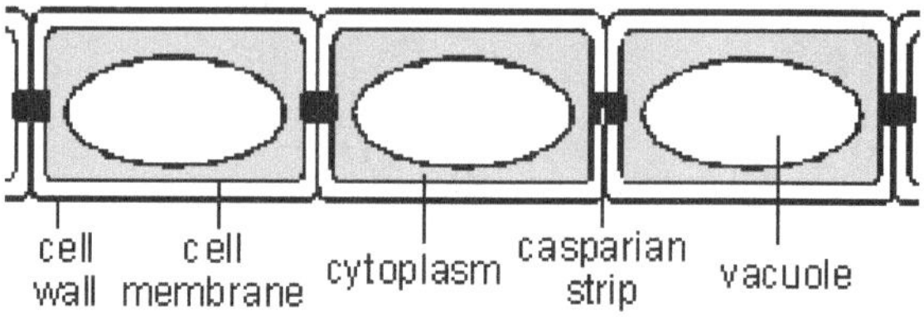

- Pericycle. A layer of undifferentiated meristematic (growing) cells.
- Vascular Tissue. This contains xylem and phloem cells, which are continuous with the stem vascular bundles. The arrangement is different, and the xylem usually forms a star shape with 2-6 arms.

Xylem Tissue

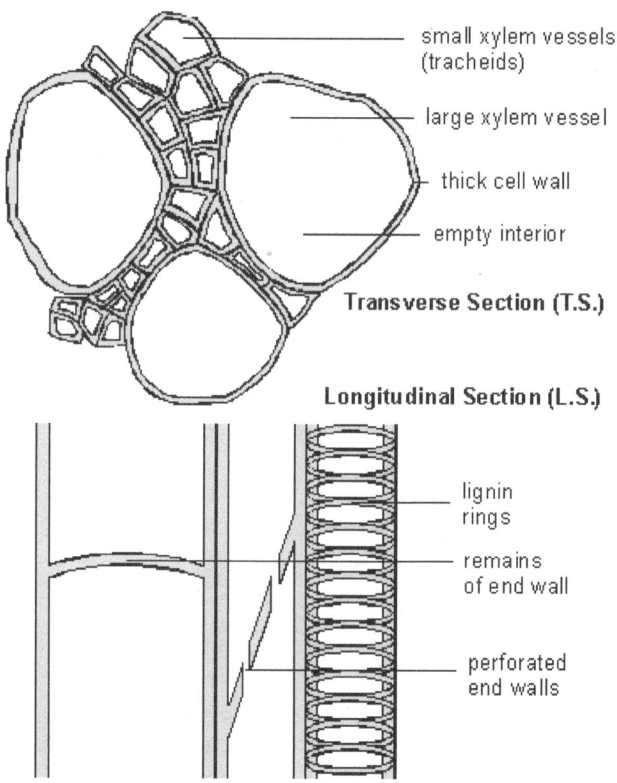

Xylem tissue is composed of dead cells joined together to form long empty tubes. Different kinds of cells form wide and narrow tubes, and the end cells walls are either full of holes, or are absent completely. Before death the cells form thick cell walls containing lignin, which is often laid down in rings or helices, giving these cells a very characteristic appearance under the microscope. Lignin makes the xylem vessels very strong, so that they don't collapse under pressure, and they also make woody stems strong.

Phloem Tissue

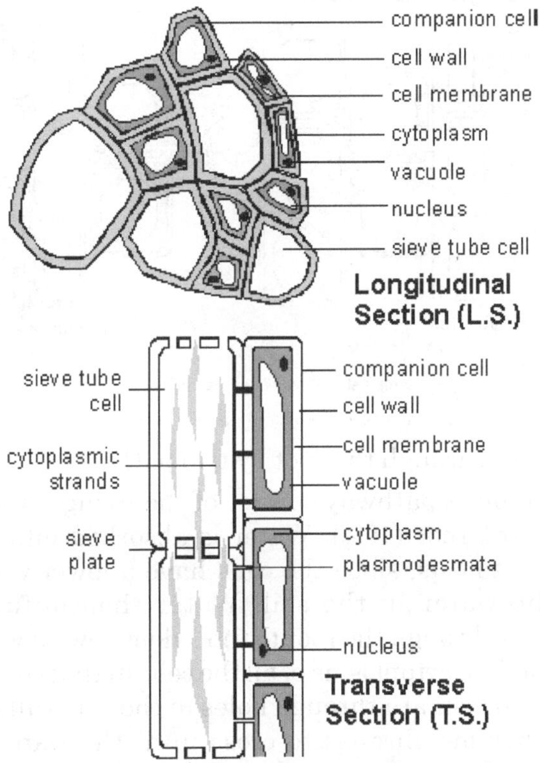

Phloem tissue is composed of sieve tube cells, which form long columns with holes in their end walls called sieve plates. These cells are alive, but they lose their nuclei and other organelles, and their cytoplasm is reduced to strands around the edge of the cells. These cytoplasmic strands pass through the holes in the sieve plates, so forming continuous filaments. The centre of these tubes is empty. Each sieve tube cell is associated with one or more companion cells, normal cells with nuclei and organelles. These companion cells are connected to the sieve tube cells by plasmodesmata, and provide them with proteins, ATP and other nutrients.

Water Transport in Plants

Vast amounts of water pass through plants. A large tree can use water at a rate of 1 dm^3 min^{-1}. Only 1% of this water is used by the plant cells for photosynthesis and turgor, and the remaining 99% evaporates from the leaves and is lost to the atmosphere. This evaporation from leaves is called transpiration.

The movement of water through a plant can be split into three sections: through the roots, stem and leaves:

Movement through the Roots

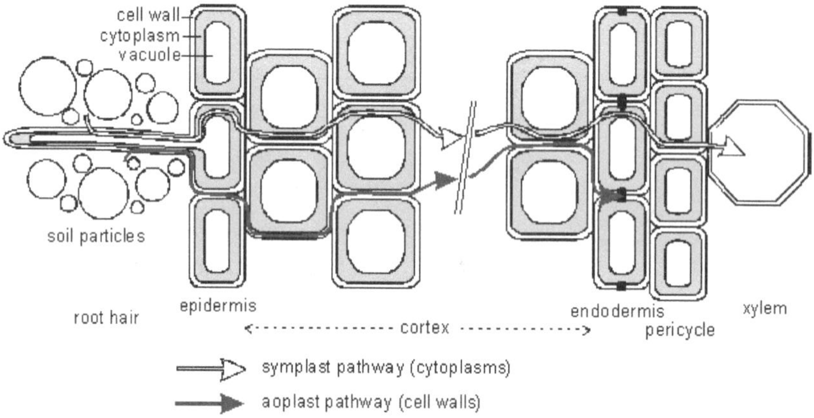

Water moves through the root by two paths:

- The Symplast pathway consist of the living cytoplasms of the cells in the root (10%). Water is absorbed into the root hair cells by osmosis, since the cells have a lower water potential that the water in the soil. Water then diffuses from the epidermis through the root to the xylem down a water potential gradient. The cytoplasms of all the cells in the root are connected by plasmodesmata through holes in the cell walls, so there are no further membranes to cross until the water reaches the xylem, and so no further osmosis.
- The Apoplast pathway consists of the cell walls between cells (90%). The cell walls are quite thick and very open, so water can easily diffuse through cell walls without having to cross any cell membranes by osmosis. However the apoplast pathway stops at the endodermis because of the waterproof casparian strip, which seals the cell walls. At this point water has to cross the cell membrane by osmosis and enter the symplast. This allows the plant to have some control over the uptake of water into the xylem.

The uptake of water by osmosis actually produces a force that pushes water up the xylem. This force is called root pressure, which can be measured by placing a manometer over a cut stem, and is of the order of 100 kPa (about 1 atmosphere). This helps to push the water a few centimetres up short and young stems, but is nowhere near enough pressure to force water up a long stem or a tree. Root pressure is the cause of guttation, sometimes seen on wet mornings, when drops of water are forced out of the ends of leaves.

Movement through the Stem

The xylem vessels form continuous pipes from the roots to the leaves. Water can move up through these pipes at a rate of 8m h^{-1}, and can reach a height of over 100m. Since the xylem vessels are dead, open tubes, no osmosis can occur within them.

The driving force for the movement is transpiration in the leaves. This causes low pressure in the leaves, so water is sucked up the stem to replace the lost water. The column of water in the xylem vessels is therefore under tension (a stretching force). Fortunately water has a high tensile strength due to the tendency of water molecules to stick together by hydrogen bonding (cohesion), so the water column does not break under the tension force. This mechanism of pulling water up a stem is sometimes called the cohesion-tension mechanism.

The very strong lignin walls of the xylem vessels stops them collapsing under the suction pressure, but in fact the xylem vessels (and even whole stems and trunks) do shrink slightly during the day when transpiration is maximum.

Movement through the Leaves

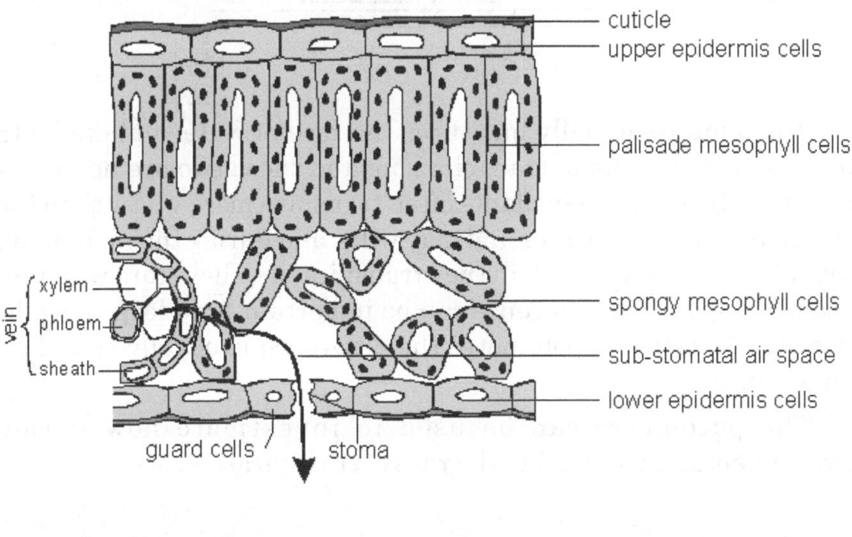

The xylem vessels ramify in the leaves to form a branching system of fine vessels called leaf veins. Water diffuses from the xylem vessels in the veins through the adjacent cells down its water potential gradient. As in the roots, it uses the symplast pathway through the living cytoplasm and the apoplast pathway through the non-living cell walls. Water evaporates from the spongy cells into the sub-stomatal air space, and diffuses out through the stomata.

Evaporation is endothermic and is driven by solar energy, which is therefore the ultimate source of energy for all the water movements in plants:

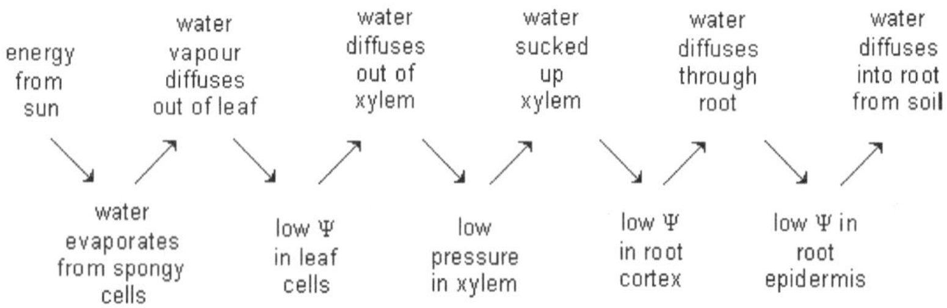

Factors Affecting Transpiration

The rate of transpiration can be measured in the lab using a potometre ("drinking metre"):

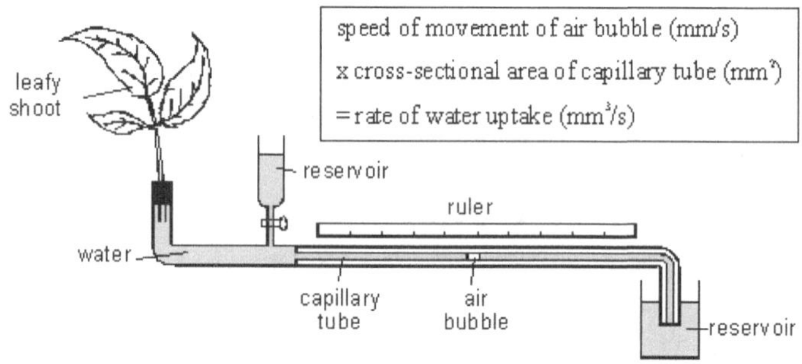

A potometre actually measures the rate of water uptake by the cut stem, not the rate of transpiration; and these two are not always the same. During the day plants often transpire more water than they take up (i.e. they lose water and may wilt), and during the night plants may take up more water than they transpire (i.e. they store water and become turgid). The difference can be important for a large tree, but for a small shoot in a potometre the difference is usually trivial and can be ignored.

The potometre can be used to investigate how various environmental factors affect the rate of transpiration.

- Light. Light stimulates the stomata to open allowing gas exchange for photosynthesis, and as a side effect this also increases transpiration. This is a problem for some plants as they may lose water during the day and wilt.
- Temperature. High temperature increases the rate of evaporation of water from the spongy cells, and reduces air humidity, so transpiration increases.
- Humidity. High humidity means a higher water potential in the air, so a lower water potential gradient between the leaf and the air, so less evaporation.
- Air movements. Wind blows away saturated air from around stomata, replacing it with drier air, so increasing the water potential gradient and increasing transpiration.

Many plants are able to control their stomata, and if they are losing too much water and their cells are wilting, they can close their stomata, reducing transpiration and water loss.

So long periods of light, heat, or dry air could result in a decrease in transpiration when the stomata close.

Adaptations to Dry Habitats

Plants in different habitats are adapted to cope with different problems of water availability.

Mesophytes	plants adapted to a habitat with adequate water
Xerophytes	plants adapted to a dry habitat
Halophytes	plants adapted to a salty habitat
Hydrophytes	plants adapted to a freshwater habitat

Some adaptations of xerophytes are:

Adaptation	*How it works*	*Example*
thick cuticle	stops uncontrolled evaporation through leaf cells	most dicots
small leaf surface area	less area for evaporation	conifer needles, cactus spines
low stomata density	fewer gaps in leaves	
stomata on lower surface of leaf only	more humid air on lower surface, so less evaporation	most dicots
shedding leaves in dry/cold season	reduce water loss at certain times of year	deciduous plants
sunken stomata	maintains humid air around stomata	marram grass, pine
stomatal hairs	maintains humid air around stomata	marram grass, couch grass
folded leaves	maintains humid air around stomata	marram grass,
succulent leaves and stem	stores water	cacti
extensive roots	maximise water uptake	cacti

Mineral Ion Transport in Plants

Ions are absorbed from the soil by both passive and active transport. Specific ion pumps in the membranes of root hair cells pump ions from the soil into the cytoplasms of the epidermis cells.

Two lines of evidence indicate that active transport is being used:

- The concentrations of ions inside root cells are up to 100 time greater than in the soil, so they are being transported up their concentration gradient.
- If respiratory inhibitors such as cyanide are applied to living roots, ion uptake is greatly reduced, since there is no ATP being made to drive the membrane pumps. Any remaining uptake must be passive.

The active uptake of ions is partly responsible for the water potential gradient in roots, and therefore for the uptake of water by osmosis.

Ions diffuse down their concentration gradient from the epidermis to the xylem. They travel up the xylem by mass flow as the water is pulled up the stem (in other words they are simply carried up in the flow of the xylem solution). In the leaves they are selectively absorbed into the surrounding cells by membrane pumps.

Solute Transport in Plants

The phloem contains a very concentrated solution of dissolved solutes, mainly sucrose, but also other sugars, amino acids, and other metabolites. This solution is called the sap, and the transport of solutes in the phloem is called translocation.

Unlike the water in the xylem, the contents of the phloem can move both up or down a plant stem, often simultaneously. It helps to identify where the sugar is being transported from (the source), and where to (the sink).

- During the summer sugar is mostly transported from the leaves, where it is made by photosynthesis (the source) to the roots, where it is stored (the sink).
- During the spring, sugar is often transported from the underground root store (the source) to the growing leaf buds (the sink).
- Flowers and young buds are not photosynthetic, so sugars can also be transported from leaves or roots (the source) to flowers or buds (sinks).

Surprisingly, the exact mechanism of sugar transport in the phloem is not known, but it is certainly far too fast to be simple diffusion. The main mechanism is thought to be the mass flow of fluid up the xylem and down the phloem, carrying dissolved solutes with it. Plants don't have hearts, so the mass flow is driven by a combination of active transport (energy from ATP) and evaporation (energy from the sun). This is called the mass flow theory, and it works like this:

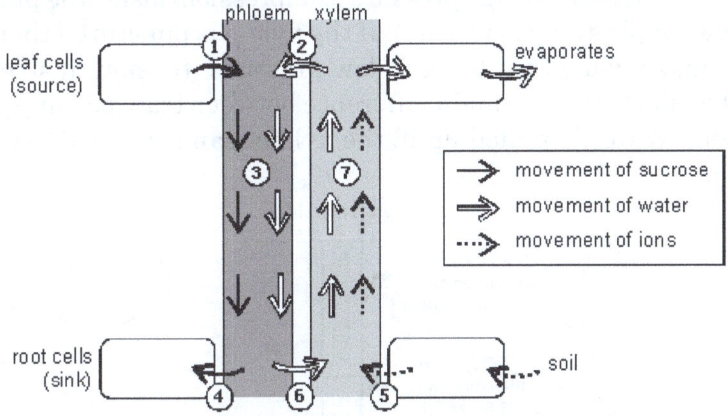

1. Sucrose produced by photosynthesis is actively pumped into the phloem vessels by the companion cells.
2. This decreases the water potential in the leaf phloem, so water diffuses from the neighbouring xylem vessels by osmosis.
3. This is increases the hydrostatic pressure in the phloem, so water and dissolved solutes are forced downwards to relieve the pressure. This is mass flow: the flow of water together with its dissolved solutes due to a force.
4. In the roots the solutes are removed from the phloem by active transport into the cells of the root.
5. At the same time, ions are being pumped into the xylem from the soil by active transport, reducing the water potential in the xylem.
6. The xylem now has a lower water potential than the phloem, so water diffuses by osmosis from the phloem to the xylem.
7. Water and its dissolved ions are pulled up the xylem by tension from the leaves. This is also mass flow.

This mass-flow certainly occurs, and it explains the fast speed of solute translocation. However there must be additional processes, since mass flow does not explain how different solutes can move at different speeds or even in different directions in the phloem. One significant process is cytoplasmic streaming: the active transport of molecules and small organelles around cells on the cytoskeleton.

Translocation Experiments

Puncture Experiments

If the phloem is punctured with a hollow tube then the sap oozes out, showing that there is high pressure (compression) inside the phloem (this is how maple syrup is tapped). If the xylem is punctured then air is sucked in, showing that there is low pressure (tension) inside the xylem. This illustrates the main difference between transport in xylem and phloem: Water is pulled up in the xylem, sap is pushed down in the phloem.

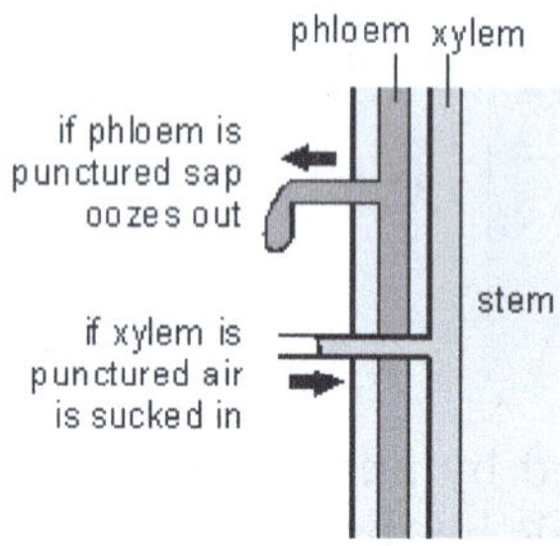

Ringing Experiments

Since the phloem vessels are outside the xylem vessels, they can be selectively removed by cutting a ring in a stem just deep enough to cut the phloem but not the xylem. After a week there is a swelling above the ring, reduced growth below the ring and the leaves are unaffected. This was early evidence that sugars were transported downwards in the phloem.

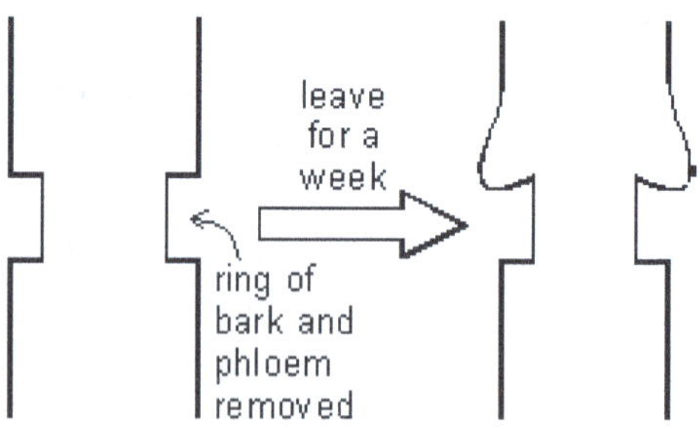

Radioactive Tracer Experiments

Radioactive isotopes can be used trace precisely where different compounds are being transported from and to, as well as measuring the rate of transport. The radioactivity can be traced using photographic film (an autoradiograph) or a GM tube. This techniques can be used to trace sugars, ions or even water.

In a typical experiment a plant is grown in the lab and one leaf is exposed for a short time to carbon dioxide containing the radioactive isotope ^{14}C. This $^{14}CO_2$ will be taken up by photosynthesis and the ^{14}C incorporated into glucose and then sucrose. The plant is then frozen in liquid nitrogen to kill and fix it quickly, and placed onto photographic film in the dark. The resulting autoradiograph shows the location of compounds containing ^{14}C.

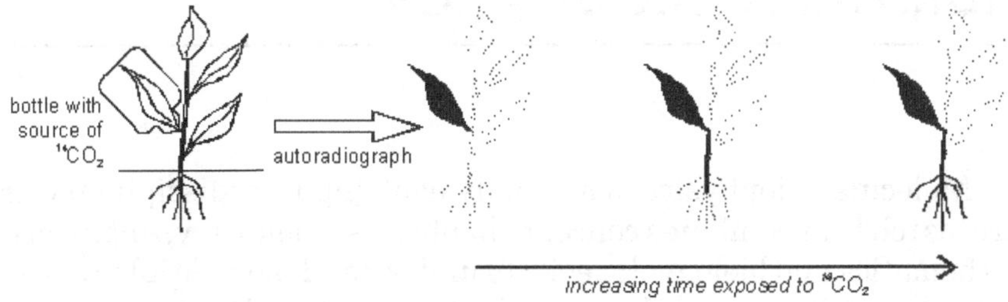

This experiment shows that organic compounds (presumably sugars) are transported downwards from the leaf to the roots. More sophisticated experiments using fluorescently labelled compounds can locate the compound specifically to the phloem cells.

Aphid Stylet Experiments

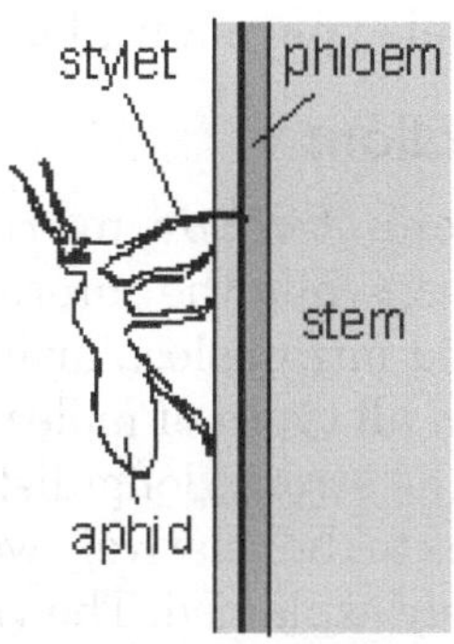

Aphids, such as greenfly, have specialised mouthparts called stylets, which they use to penetrate phloem tubes and sup of the sugary sap therein. If the aphids are anaesthetised with carbon dioxide and cut off, the stylet remains in the phloem so pure phloem sap can be collected through the stylet for analysis. This surprising technique is more accurate than a human with a syringe and the aphid's enzymes ensure that the stylet doesn't get blocked.

Chapter 4

Molecular Biophysics

Molecular biophysics is a rapidly evolving interdisciplinary area of research that combines concepts in physics, chemistry, engineering, mathematics and biology. It seeks to understand biomolecular systems and explain biological function in terms of molecular structure, structural organization, and dynamic behaviour at various levels of complexity (from single molecules to supramolecular structures, viruses and small living systems). The technical challenges are formidable, and the discipline has required development of specialized equipment and procedures capable of imaging and manipulating minute living structures, as well as novel experimental approaches.

Inter-Molecular Interactions

Interactions between two or more molecules are called intermolecular interactions, while the interactions between the atoms within a molecule are called intramolecular interactions. Intermolecular interactions occur between all types of molecules or ions in all states of matter. They range from the strong, long-distance electrical attractions and repulsions between ions to the relatively weak dispersion forces which have not yet been completely explained. The various types of interactions are classified as (in order of decreasing strength of the interactions):

ion - ion

ion - dipole

dipole - dipole

ion - induced dipole

dipole - induced dipole

dispersion forces

Without these interactions, the condensed forms of matter (liquids and solids) would not exist except at extremely low temperatures. We will explore these various forces and interactions in the gas phase to understand why some materials vaporize at very low temperatures, and others persist as solids or liquids to extremely high temperatures.

Covalent interactions (bonds) hold atoms together within molecules. Covalent bonds are strong, and their enthalpies are on the order of 100 kcal/mole (400 kjoule/mole). Covalent bonds remain intact when ice melts, when water boils, when proteins unfold, when RNA unfolds, when DNA strands separate, and when membranes disassemble.

The processes of melting, boiling, and unfolding involve disruption of *molecular interactions* (also called noncovalent interactions or intermolecular interactions), which are interactions *between* molecules. Molecular interactions are important in diverse fields of protein folding, drug design, separation technologies, etc.

Molecular interactions seem weak. The enthalpy of a given molecular interaction is 1-10 kcal/mol, which in the lower limit is on the order of RT and in the upper limit is significantly less than a covalent bond. Even though they are weak individually, cumulatively the energies of molecular interactions are significant.

Boiling Points. Molecular interactions underlie the differences in boiling points of liquids. The boiling point of H_2O is hundreds of degrees greater than the boiling point of N_2 because of differences in molecular interactions in H_2O(liq) versus N_2(liq). The forces between molecules in H_2O(liq) are greater than those in N_2(liq). When a molecule transitions from the liquid to the gas phase (as during boiling), all molecular interactions are broken. In an ideal gas there are no adhesive molecular interactions. It is a good general rule that differences in boiling temperatures give good qualitative estimates of strengths of molecular interactions within various liquids.

Pushing and Pulling. In their native states, a protein is generally folded into a globular structure and DNA is double-stranded. The number of molecular interactions within a folded native protein or a long double-stranded DNA is enormous. However if you unfold the protein (denature it) or separate the DNA strands, the interior becomes exposes to the surrounding aqueous media. Intra-molecular interactions in the native state are replaced by molecular interactions with water molecules in the non-native state.

Huge numbers of molecular interactions within a native state are balanced by huge numbers of molecular interactions of the non-native state (with surrounding water molecules). Native states are held in 'delicate balance between powerful countervailing forces' contributed primarily by massive numbers of molecular interactions.

Native states are marginally stable: the equilibrium constant for the transition from native to unfolded state is small.

A small difference between these large effects, can be contributed by weak interactions, changing the direction of the folding 'reaction'. A small change in pH or temperature can tip the balance.

Have you ever denatured a protein (converted it from the native folded state to a non-native unfolded state also known as a random coil)? Yes. When you heat an egg to around 60° C, the albumin proteins denature and aggregate.

The aggregated protein forms large assemblies that scatter light, giving the egg a white appearance. When you add lemon juice to milk, the pH drops and the proteins denature and aggregate. Have you ever melted DNA? Yes, if you have run a PCR reaction.

Molecular interactions were discovered by the Dutch scientist Johannes Diderik van der Waals. He noticed that molecules are sticky. The phrase 'van der Waals interaction' has come to mean a adhesive/ cohesive force between molecules that are close together in space. The term 'van der Waals interaction' is not sufficiently informative or descriptive for our purposes here.

A van der Waals interaction describes a totality of various types molecular interactions. We avoid the term "van der Waals interaction" because that phrase does not decompose the interactions in a physically meaningful way or provide sufficiently predictive models.

There are many different ways of parsing or classifying molecular interactions. The categories in the Table of Contents are used here because they are the clearest and easiest to understand and are broadly used in the literature.

All molecular interactions are fundamentally electrostatic in nature and can described by some variation of Coulombs Law.

Short Range Repulsion

Atoms take space. When two atoms approach each each other, at some distance the overlap of the occupied orbitals causes electrostatic repulsion between the electrons of the two atoms.

This repulsive energy between atoms acts over a very short range, but goes up sharply when that range is violated.

Atom Name	*VDW Radius (r, Å)*
H	1.2
C	1.7
N	1.6
O	1.5
P	1.8
S	1.8

The repulsion goes up as $1/R^{12}$. It is important only when atoms are in very close proximity, but then it becomes very important. Because this repulsive term rises so sharply as distance decreases it is very often reasonable to think of atoms as hard spheres, like small pool balls, described by well-defined van der Waal radii (r) and hard van der Waal surfaces.

As two atoms approach each other their van der Waals surfaces make contact when the distance between them decreases to the sum of their van der Waals radii.

At this point the repulsive energy skyrockets. The smallest distance between two non-bonded atoms is the sum of the van der Waals radii of the two atoms. A sulfur atom and a carbon atom can come no closer together than $r_S + r_C = 1.8 + 1.7 = 3.5$ Å. Of course we are assuming here that bonds do not form. When two atoms form a bond, they come close together such that van der Waals radii and surfaces are violated.

Short range repulsion is important to you. It prevents atoms from collapsing into tightly packed states of density of 10^{14} g/ml, which is the density of condensed atomic nuclei. Very high gravitation forces, as in neutron stars, overwhelm short range repulsion and cause atoms to collapse.

Here in earth, with our modest gravity, the van der Waals radius of carbon (r_C) is evident from the spacing between the layers in graphite. Those coordinates are here. The distance between atoms in different layers of graphite is never less than twice the van der Waals radius of carbon ($2 \times r_C = 2 \times 1.7 = 3.4$ Å).

The atoms within a graphite layer are covalently linked (bonded), which causes interpenetration of van der Waals surfaces. Carbon atoms within a layer are separated by 1.42 Å, which is much less than twice the van der Waals radius of carbon.

As explained in other sections of this document vdw surfaces are also violated when molecules form hydrogen bonds.

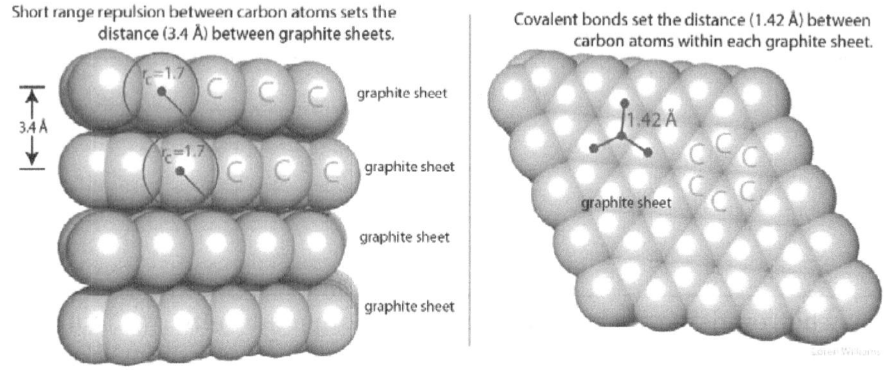

Figure: *Figure shows how short range repulsion sets the distance of 3.4 Å between sheets in graphite. If two non-bonded atoms are separated by a distance of less than the sum of their VDW radii, short range repulsion forces them apart.*

In B-DNA, the helical rise per base pair of 3.4 Å is determined by short range repulsion between the stacked bases. Bases stack on each other with spacing of 3.4 Å.

Electrostatic Interactions

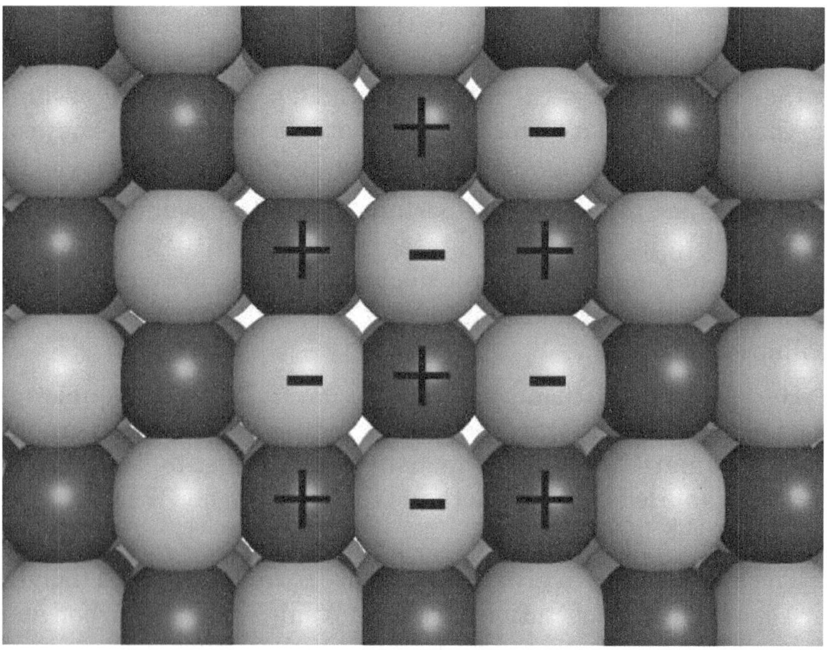

Figure: *Figure shows a cross section of a NaCl crystal. Each sodium cation experiences strong electrostatic interactions with adjacent chloride anions. In reality the ionic radius of a sodium cation is less than that of a chloride anion. The coordinates of sodium chloride are here .*

Electrostatic interactions are between and among cations and anions, which are species with formal charge of ...-2,-1,+1,+2... Electrostatic interactions can be either attractive or repulsive, depending on the signs of the charges. Favourable electrostatic interactions cause the vapor pressure of sodium chloride and other salts to be very low. If you leave table salt (NaCl) out on a table, how long does it take before it sublimes away? A very very long time. The electrostatic interactions within a sodium chloride crystal are called ionic bonds. But when a single cation and a single anion are close together, within a protein, or within a folded RNA, those interactions are considered to be non-covalent electrostatic interactions. Electrostatic interactions can be very strong, and fall off gradually with distance (1/r). As explained later in this document, electrostatic interactions are highly attenuated (dampened) by water.

Electrostatic interactions are the primary stabilizing interaction between phosphate oxygens of RNA (formal charge -1) and magnesium ions (formal charge +2), as shown in the figure below. There are many magnesium ions associated with RNA and DNA molecules *in vivo*.

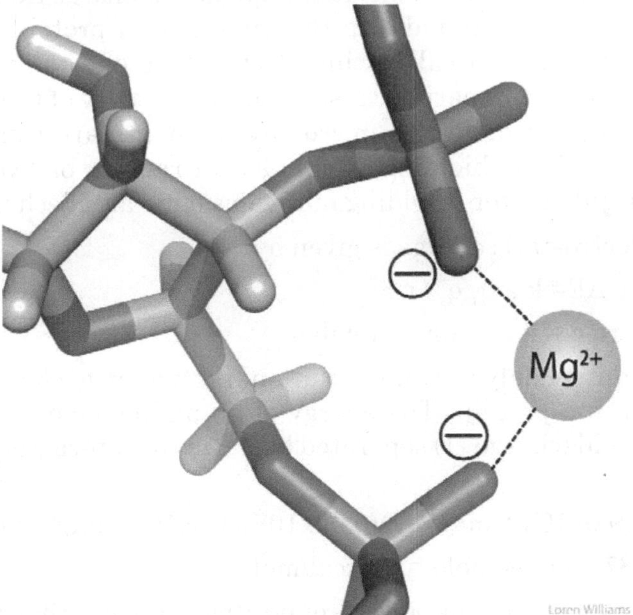

Figure: *Figure shows RNA (within the ribosome) in which anionic phosphate oxygens (-1 charge) engage in attractive electrostatic interactions with a dicationic magnesium ion (+2 charge). The O to Mg^{2+} distance is 2.1 Å. Here the dashed lines represent favourable electrostatic interactions (not hydrogen bonds: there are no hydrogen atoms between the phosphate oxygens and the Mg^{2+} ion.)*

Ion Pairs in Proteins

Favourable electrostatic interactions between paired anionic and cationic amino acid sidechains are reasonably frequent in proteins. Ion Pairs, sometimes called Salt Bridges, are formed when the charged group of a cationic amino acid (like lysine or arginine) is around 3.5 to 5.0 Å from the charged group of an anionic amino acid (like aspartate or glutamate). The charged groups in an Ion Pair are generally linked by hydrogen bonds, in addition to electrostatic interactions.

The electrostatic force between two point formal charges is given by:

$$\text{Force} = k\, q_1\, q_2 / \varepsilon\, r^2$$

where $k = 9.0 \times 10^9$ nt-metre^2 / coul^2

$q = -1.6 \times 10^{-19}$ coulombs for an electron.

r = distance between the point charges (metres)

ε = the dielectric constant of the medium (unitless).

ε reflects the tendency of the medium to shield one charge from another. ε is 1 in a vacuum, around 4 in the interior of a protein and 80 in water. The problem of calculating electrostatic effects in biological systems is complex in part because of non-uniformity of the dielectric environment. The dielectric micro-environments are complex and variable, with less shielding of charges in regions of hydrocarbon sidechains and greater shielding in regions of polar sidechains.

The electrostatic energy is given by:

$$\Delta E = k\, a\, q_1\, q_2 / \varepsilon\, r$$

where a = Avogadro's number.

One can crudely estimate the energetics of a charge-charge interaction in a protein. The energy of an amine (charge +1) and a carboxylic acid (charge -1) separated by 4 E in the interior of protein is given by:

$$\Delta E = -(9.0\text{x}10^9 \text{nt-m}^2/\text{coul}^2)(6.02\text{x}10^{23})(1.6\text{x}10^{-19}\text{coul})^2 / 4(\,4\text{x}10^{-10}\text{m})$$

$$= 87 \text{ kjoules/mole} = 21 \text{ kcal/mole}$$

This rough approximation is around 10-fold greater than the values determined experimentally. An ion pair contributes 1-4 kcal/mol (favourable ΔG) to the stability of a native protein.

A note on nomenclature. The attractive forces between a Mg^{2+} ion and phosphate groups (above) are called electrostatic interactions. This label is unfortunate because ALL molecular interactions are inherently electrostatic in nature. It might have been better to have called these charge-charge interactions.

However, by convention we use the term electrostatic to describe interactions between formally charged species. We use other terms (dipole-dipole, dispersion...) to describe (electrostatic) interactions between partial charges.

Dipole Interactions

Molecular dipoles: Before we can think very much about dipole-dipole interactions, or understand them, we have to know about dipoles and electronegativities. In a molecule with unlike atoms, electrons are not shared equally.

The tendency of any atom to pull electrons towards itself, and away from other atoms, is characterized by a quantity called electronegativity. A greater difference in the electronegativities of two bonded atoms causes the bond between them to become more polar, and the partial charges on the atoms to become larger in magnitude.

In biological systems, oxygen is generally the most electronegative atom, carrying the largest partial negative charge.

Atom Name	*Electronegativity*
H	2.2
C	2.6
N	3.0
O	3.5
P	2.2
S	2.5

In a molecule composed of atoms of various electronegativities, the atoms with lowest (smallest) electronegativities hold partial positive charges and the atoms with the greatest electronegativities hold partial negative charges.

In a methanol molecule (CH_3OH), the electronegative oxygen atom pulls electron density away from the carbon and hydrogen atoms. In a water molecule, the electronegative oxygen atom pulls electron density away from the hydrogen atoms. The oxygen atom of water carries a partial negative charge. The hydrogen atoms carry partial positive charges. This phenomena of charge separation is called polarity. Water is a polar molecule. N_2 is a non-polar molecule because the two nitrogen atoms share electrons equally.

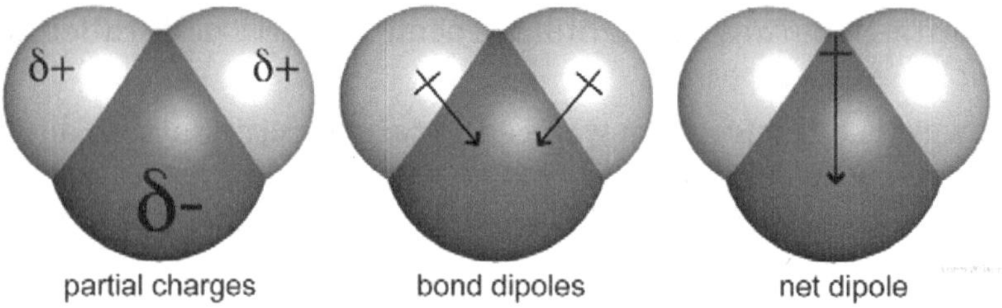

Figure: *Figure shows the charge distribution and dipole moment of a water molecule. The electronegative oxygen atom pulls electron density away from the hydrogen atoms. The oxygen carries a partial negative charge and the hydrogen atoms carry partial positive charges. Bond dipoles (center) and molecular dipoles (right) can be represented as vectors. The arrows point from positive charge toward negative charge.*

The extent of charge separation within a molecule is characterized by the dipole moment μ. The dipole moment of a molecule is determined by the magnitudes of the partial charges and by the distances between them. To quantitate dipole moments, charges are expressed in esu's and distances in centimetres.

The dipole moment of an electron and a proton separated by 1 Å equals:

$$(4.8 \text{ x } 10^{-10} \text{ esu})\ (10^{-8}\text{cm}) = 4.8 \text{ x } 10^{-18} \text{ esu cm}$$
$$= 4.8 \text{ Debye}$$

The dipole moment of water is 1.85 Debye (HCl = 1.1 D; CH_3Cl = 1.9 D; HCN = 2.9 D; NH_3 = 1.47.

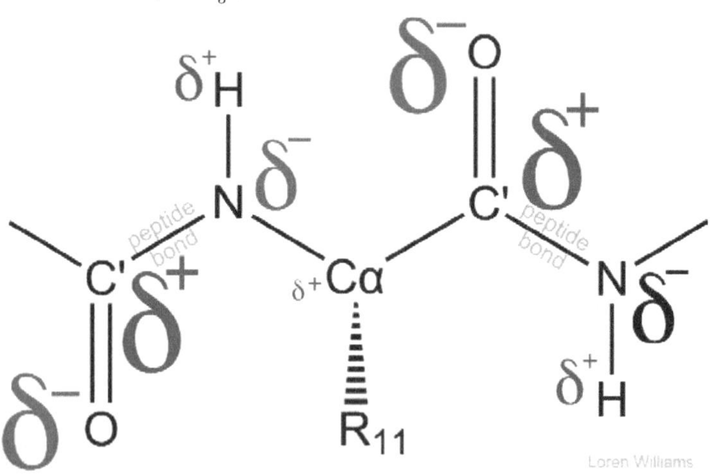

Figure: *Figure shows the partial charges within a polypeptide. The symbol size (d) is scaled to the magnitude of the partial charge. Oxygen atoms are the most electronegative and have the greatest negative partial charge.*

Partial Charges on the Atoms of a Peptide

Atom Name	*partial charge (e⁻)*
N	-0.42
HN	+0.27
Ca	+0.12
C'	+0.60
O	-0.57

The orientation of the dipole moment of a peptide is approximately parallel to the N-H bond and in magnitude is around 3.7 Debye.

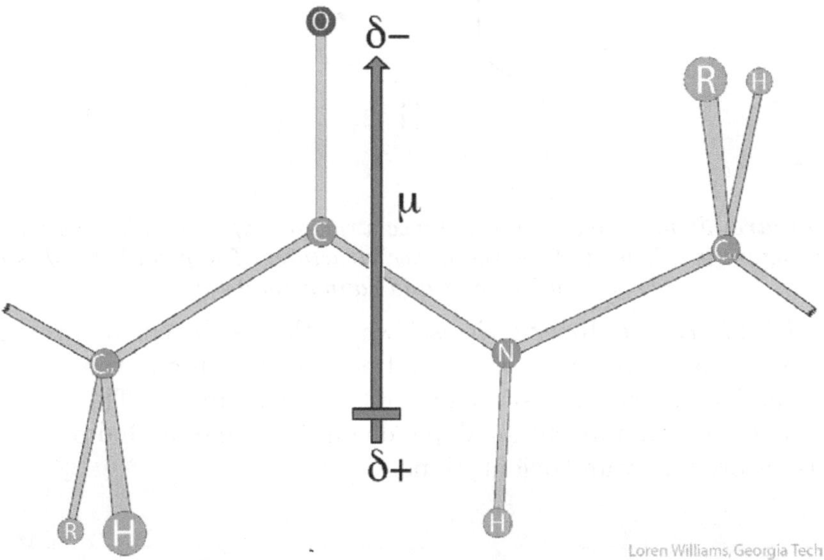

Figure: *Figure shows the orientation of the dipole moment of a peptide.*

The large dipole moment of a peptide bond should lead one to expect that dipolar interactions are important in protein conformation and interactions. They are the large dipole of a peptide bond can be attributed in part to resonance.

However, these simple resonance structures present an oversimplified view of the electronic structure of a peptide.

A dipole can interact with point charges. These interactions are called Charge-Dipole Interactions (or Ion-Dipole Interactions). Dipoles can also interact with other dipoles (called Dipole-Dipole Interactions), and can induce charge distribution (polarization) in surrounding molecules (causing Dipole-Induced Dipole Interactions). We will discuss each of these interactions separately.

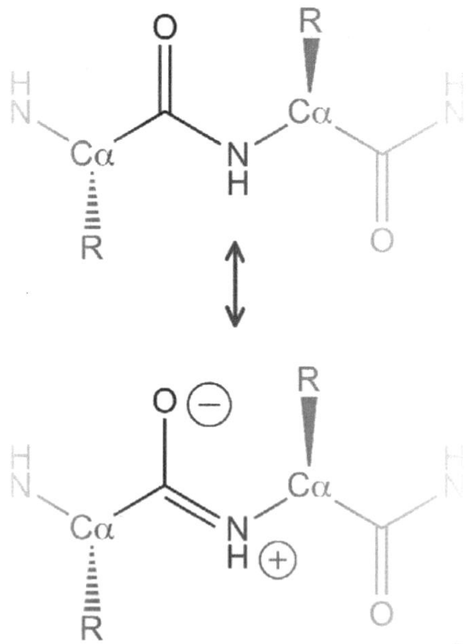

***Figure:** Figure shows two resonance structures of a peptide. The real structure is a hybrid of the resonance structures. The peptide bond is a partial double bond and cannot rotate freely.*

Dipole-Dipole Interactions: Two dipoles 'feel' each other through space. The positive end of the first dipole is attracted to the negative end and is repelled by positive end of the second dipole. The strength of a dipole-dipole interaction depends on the size of both dipoles and on their proximity and orientations.

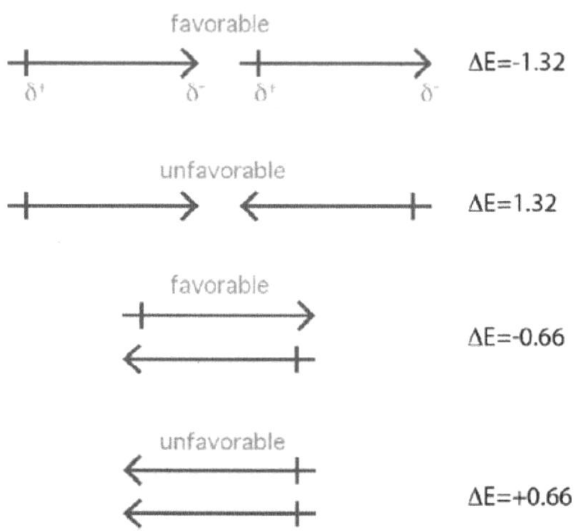

***Figure:** Figure shows how dipole-dipole interactions depend on the orientations of the dipoles. Dipole-dipole interactions can be attractive or repulsive.*

The total interaction energy between two dipoles can be either positive or negative. Parallel end to end dipoles attract while antiparallel end to end dipoles repel. Listed below are the energies of interaction for various orientations of two dipoles with moments of 1 Debye at a distance of 5 Å in a medium of $\varepsilon = 4$.

In liquids the orientations of molecular dipoles change rapidly as molecules tumble about.

However, dipole moments tend to orient favourably. Therefore, in liquid acetone for example, favourable dipole-dipole interactions outweigh unfavourable dipole-dipole interactions. Dipole-dipole interactions fall off with $1/r^3$.

Dipole-Induced Dipole Interactions: A molecule with a permanent dipole moment creates a surrounding electrostatic field. This electrostatic field will shift the electrons around a bit (induce a dipole moment) in a second molecule that is located nearby in space. This induction of a dipole moment in one molecule by another is called *polarization*. The ease with which electrons are shifted by an external electronic field is called polarizability.

Molecules with p electrons, such as phenylalanine and tryptophan, are more polarizable than molecules such as isoleucine, which lack π electrons.

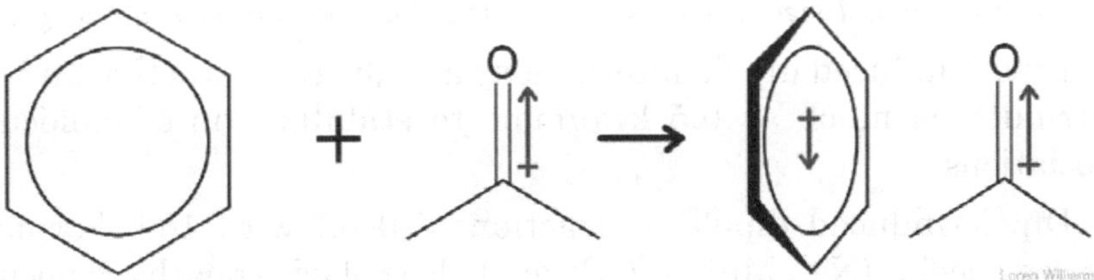

Figure: *Figure shows how a static dipole can induce a dipole in an adjacent molecule. When two isolated molecules (left) are brought together in a liquid or solid (right), the static dipole 'polarizes' the adjacent molecule. p Electrons are more polarizable (more easily perturbed by an adjacent dipole) than s electrons. The strength of a dipole-induced dipole interaction depends on the size of the dipole moment of the first molecule and on the polarizability of the second molecule.*

Dipole-induced dipole interactions are important even between two molecules that both have permanent dipoles. A permanent dipole is changed by the dipole of an adjacent molecule. For example, in liquid water (where molecules are close together), *all* water molecules are polarized. The permanent dipole of each water molecule polarizes all adjacent water molecules. The dipole of a water molecule induces a change in the dipole of a nearby water molecule, compared to the permanent dipole of an isolated water molecule (in the gas phase).

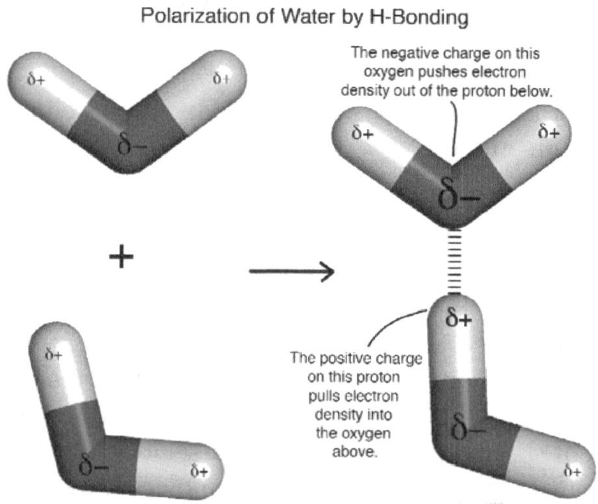

Figure: *Figure shows how water molecules polarize each other. Each water molecule polarizes neighboring water molecules and increases neighboring dipole moments. When the two water molecules approach each other and form a hydrogen bond as shown here, the partial negative charge on the oxygen of the top water molecule is increased in magnitude, and the partial positive charge on the proton of the bottom water molecule is also increased. Here the symbol size is scaled to the magnitude of the partial charge.*

Dipole-induced dipole interactions are always attractive and can contribute as much as 0.5 kcal/mole to stabilization of molecular associations.

Dipole-induced dipole interactions fall off with $1/r^4$. Formally charged species (Na^+, Mg^{2+}, $-COO^-$, etc.) also polarize nearby molecules and induce favourable dipoles.

The resulting interactions, called charge-induced dipole interactions (or ion-induced dipole interactions). These interactions are important, for example in protein structure, but are not broken out into a separate section in this document.

Dipole-Charge Interactions: A molecule with a permanent dipole can interact favourably with cations and anions. This type of interaction is called a dipole-charge or ion-dipole interaction.

Dipole-charge interactions are why sodium chloride, composed cationic sodium ions and anionic chloride ions, and other salts tend to interact well with water, and are very soluble in water, which has a strong dipole.

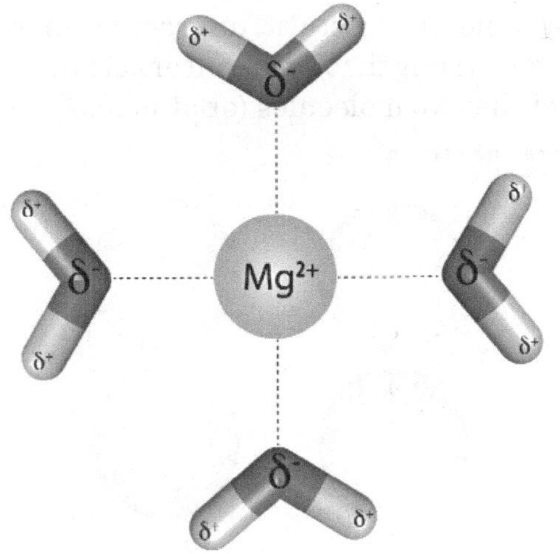

Figure: *Figure shows four water molecules interacting favorably with a magnesium dication. The negative ends of the water dipoles are directed toward the positively charged magnesium ion. If the ion is an anion, such as chloride, the water molecules switch direction and direct the positive ends of their dipoles toward the anion. Here the dashed lines do not represent hydrogen bonds. There are no hydrogen atoms between the Mg2+ cation and the water oxygen atoms.*

Fluctuating Dipoles (Dispersive Interactions, London Forces)

We can see resonance all around us. A child on a swing, the tides in the Bay of Fundy and the strings on a violin all illustrate natural resonant frequencies of physical systems. The Tacoma Narrows Bridge is one of the most famous examples of resonance.

Molecules resonate too. Electrons, even in a spherical atom like Helium or Xenon, fluctuate over time according to the natural resonant frequency of that atom. Even though chemists describe atoms like Helium and Xenon as spherical, if you could take a truly instantaneous snapshot of one of these atoms, you would always catch it in a transient non-spherical state. Xenon is spherical on average, but not at any instantaneous timepoint.

Fluctuating electrons cause fluctuating dipoles. When the instantaneous electron density of a molecule or atom fluctuates, the dipole moment also fluctuates. Therefore, molecules and atoms behave like oscillating dipoles. In molecules that are located nearby to each other the oscillating dipoles sense each other and couple. They oscillate in synch. The movements of the electrons in adjacent molecules are correlated. Electrons in one atom tend to run away from those in the next, because of electrostatic repulsion. Coupled fluctuating dipoles experience favourable electrostatic interaction known as dispersive interactions. The strength of the interaction is related to the polarizabilities of the two molecules (or atoms).

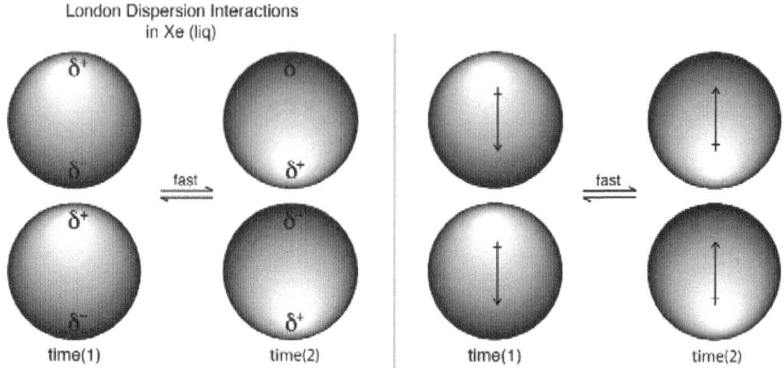

Figure: *Figure shows how fluctuating dipoles of liquid Xenon (or Helium or Neon, etc) are coupled. Darker blue indicates higher electron density. The fluctuations are correlated and are very fast, on the femtosecond (10-15 second) timescale. Adjacent Xenon atoms experience electrostatic attraction from the transient dipoles. Two different representations of fluctuating dipoles are shown.*

Dispersive interactions are always attractive and occur between any pair of molecules, polar or non-polar, that are nearby to each other. Dispersive interactions are the only attractive forces between atoms in liquid He (bp 4.5 K), Ne (27K), Ar (87K) and between molecules of N_2 (77K). Without dispersive interactions there would be no liquid state for the Nobles. About a 25% of the attractive forces between water molecules are dispersive in nature. The total number of pairwise atom-atom dispersive interactions within a folded protein is enormous, so that dispersive interactions can make a large contribute to stability.

Fluctuating dipole interactions fall off with $1/r^6$.

Hydrogen Bonding

General Description of Hydrogen Bonds: The idea that a single hydrogen atom could bond simultaneously to two other atoms was proposed in 1920 by Latimer and Rodebush and their advisor, G. N. Lewis. Maurice Huggins, who was also a student in Lewis' lab, describes the hydrogen bond in his 1919 dissertation.

An atom (A) containing a basic lone pair of electrons (i.e., a Lewis Base) can interact favourably with a hydrogen atom (H) that is covalently bound to a donor atom (D). A hydrogen bond requires that atoms A and D are both electronegative. H-D must have acidic properties. In a hydrogen bond, the Lewis Base (A) is called hydrogen bond acceptor and the acidic group (H-D) is called the hydrogen bond donor. Hydrogen is the *only* atom that (i) forms covalent bonds with electronegative atoms like N and O, *and* (ii) uses the inner shell (1S) electron(s) for bonding. When its bonding partner pulls electrons away from hydrogen, the nucleus (a proton) is exposed on the back side (distal from the bonding partner). An unshielded proton is thus available for electrostatic interactions with electron lone pairs. Hydrogen is the only atom that exposes its nucleus this way. Other atoms have inner shell non-bonding electrons that shield the nucleus. The salient features of a generic hydrogen bond are illustrated in the figure below.

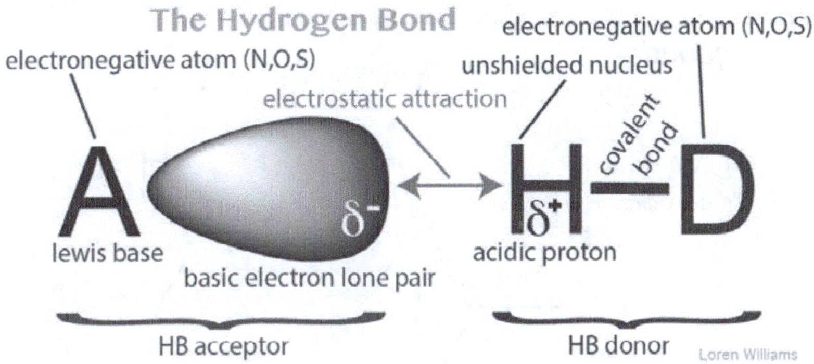

Figure: *Figure illustrates the elements of a hydrogen bond, including the HB acceptor and HB donor, the lone pair and the acidic proton. N, O, S are the predominant hydrogen bonding atoms (A & D) in biological systems.*

A hydrogen bond is *not* an acid-base reaction, where the proton (H^+) is fully transferred from H-D to A to form D^- and HA^+. However, the strength of a hydrogen bond correlates well with the acidity of donor H-D and the basicity of acceptor A. In a hydrogen bond, the H^+ is partially transfered from H-D to A, but H^+ remains covalently attached to D. The H-D bond remains intact.

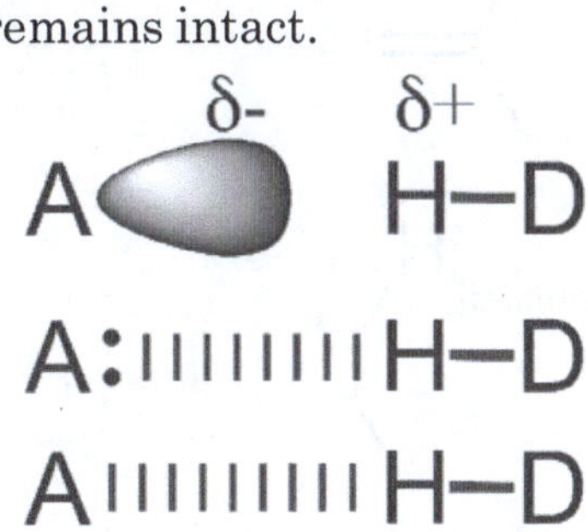

Figure: *Figure illustrates three different styles for representing a hydrogen bond. Atom A is the Lewis base (for example the N in NH3 or the O in H2O) and the atom D is electronegative (for example O, N or S). The conventional nomenclature is confusing: a hydrogen bond is not a covalent bond.*

The most common hydrogen bonds in biological systems involve oxygen and nitrogen atoms as A and D. Keto groups (=O), amines (R_3N), imines (R=N-R) and hydroxyl groups (-OH) are the most common hydrogen bond acceptors in DNA, RNA, proteins and complex carbohydrates. Hydroxyl groups and amines/imines are the most common hydrogen bond donors. Hydroxyls and amines/imines can both donate and accept hydrogen bonds.

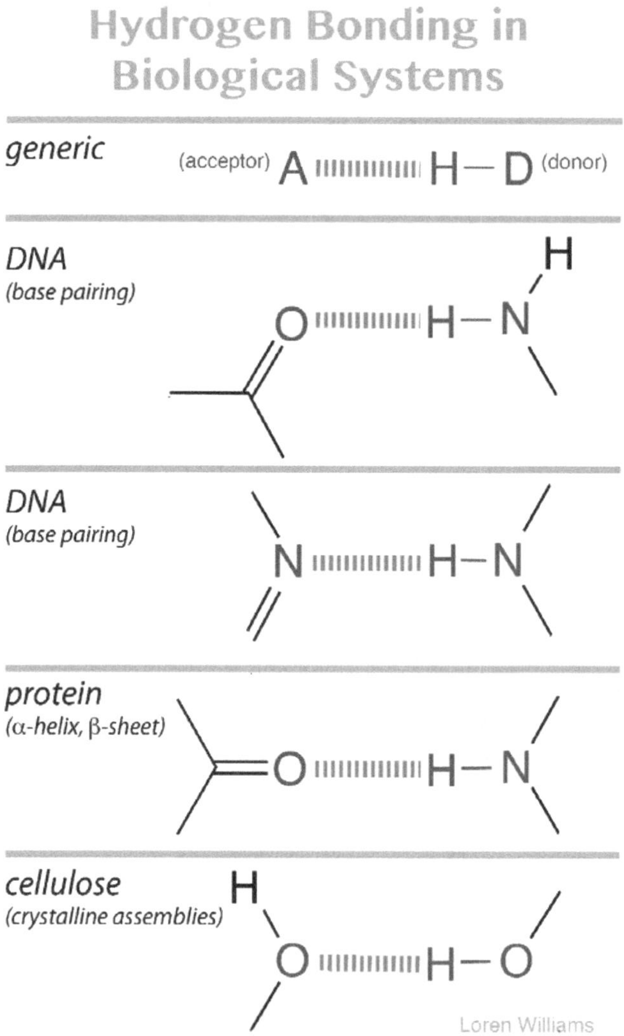

Figure: *Figure shows the most common hydrogen bond acceptors and donors in biological macromolecules.*

In traversing the Period Table, increasing the electronegativity of atom D strips electron density from the proton (in H-D), increasing its partial positive charge, and increasing the strength of any hydrogen bond. Thiols (-SH) can both donate and accept hydrogen bonds but these are generally weak, because sulfur is not sufficiently electronegative. Hydrogen bonds involving carbon, where H-D equals H-C, are observed, although these are weak and infrequent. C is insufficiently electronegative to form good hydrogen bonds. Hydrogen bonds are essentially electrostatic in nature, although the energy can be decomposed into additional contributions from polarization, exchange repulsion, charge transfer, and mixing.

Hydrogen bond strengths form a continuum. Strong hydrogen bonds of 20-40 kcal/moll, generally formed between charged donors and acceptors, are nearly as strong as covalent bonds, Weak hydrogen bonds of 1-5 kcal/mol, sometimes formed with carbon as the proton donor, are no stronger than conventional dipole-dipole interactions. Moderate hydrogen bonds, which are the most common, are formed between neutral donors and acceptors are from 3-12 kcal/mol.

Geometry of Hydrogen Bonds: The geometry of a hydrogen bond can be described by three quantities, the D to H distance, the H to A distance, and the D to H to A angle.

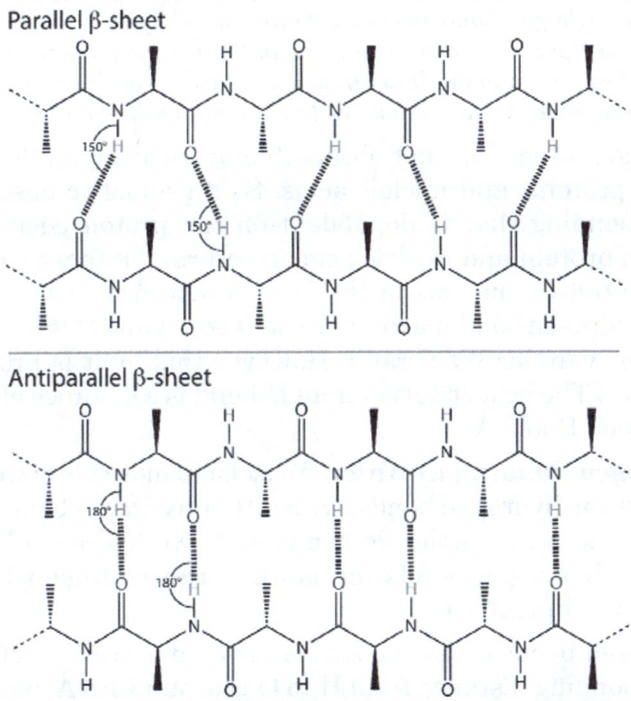

Figure: *Figure illustrates the non-linearity of parallel β-sheet hydrogen bonds and the linearity of antiparallel β-sheet hydrogen bonds.*

The distances depend on the atom types of A and D. If both A and D are oxygen atoms, then optimally, H to A = 1.8 Å and D to H = 1.0 Å. The most stable hydrogen bonds are close to linear (D to H to A angle of 180°). The hydrogen bonds in antiparallel β-sheets are linear, while the hydrogen bonds in parallel β-sheets are non-linear.

Hydrogen bonds can be two-centre (as in a β-sheets and ideal ice), three-centre, or four-centre. Two-centre hydrogen bonds are generally shorter, more linear, and stronger than three- or four-centre hydrogen bonds. Three-centre bonds are sometimes called bifurcated while four centred hydrogen bonds are sometimes called trifurcated.

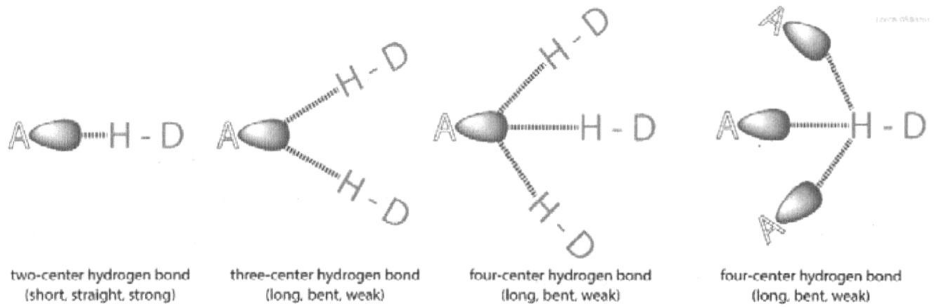

Figure: *Figure illustrates two-, three- and four-center hydrogen bonds. The two-center hydrogen bond is closest to an 'ideal' hydrogen bond, and is stronger than the other types. The left hand four-center hydrogen bonding scheme is observed in crystalline ammonium, where one acceptor lone pair has to accomodate three donors (see section on ammonia, below.*

Hydrogen atoms are not observable by x-ray crystallography as applied to proteins and nucleic acids. So a geometric description of hydrogen bonding that is dependent on the proton position is not practical in protein and nucleic acid structures. In these cases one is usually limited to analysis of the D to A distance. It is common to ascribe a hydrogen bond if a distance between A and D is less than the sum of their van der Waal radii. However this limit is probably too conservative. The best criteria for an H-bond is a distance of less than 3.4 Å between D and A.

Hydrogen Bonding in Water: An isolated molecule of water (H_2O) can form strong hydrogen bonds, with either hydrogen bond donors or acceptors. One water molecule can accept two hydrogen bonds and donate two hydrogen bonds (or more if the hydrogen bonds are bifurcated or trifurcated).

Hydrogen bonds cause violations of van der Walls surfaces. The hydrogen-bonding distance from H to O is around 1.8 Å, which is less than the sum of the O and H van der Waals radii (r_O=1.5 Å; r_H= 1.0 Å). Also notice that the hydrogen-bonding distance from O to O is around

2.8 Å, which is less than twice the van der Waals radius of oxygen (r_O=1.5 Å).

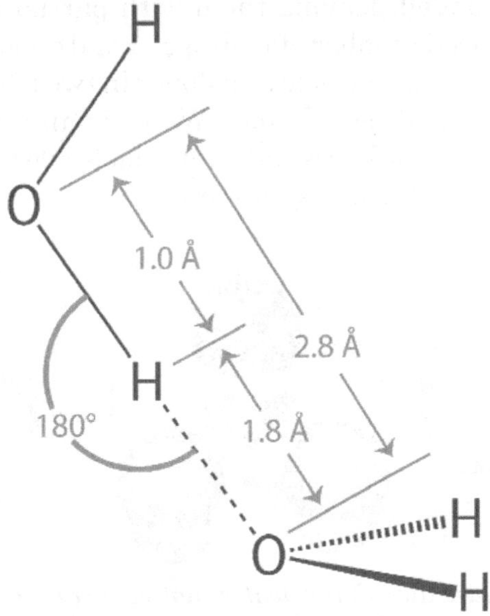

Figure: *Figure Illustrates hydrogen bonding between two water molecules as observed in crystalline water (ice). The hydrogen bonds are short, linear and strong. These are two-center hydrogen bonds. Although each water molecule in ice forms four hydrogen bonds, only one hydrogen bond is shown here.*

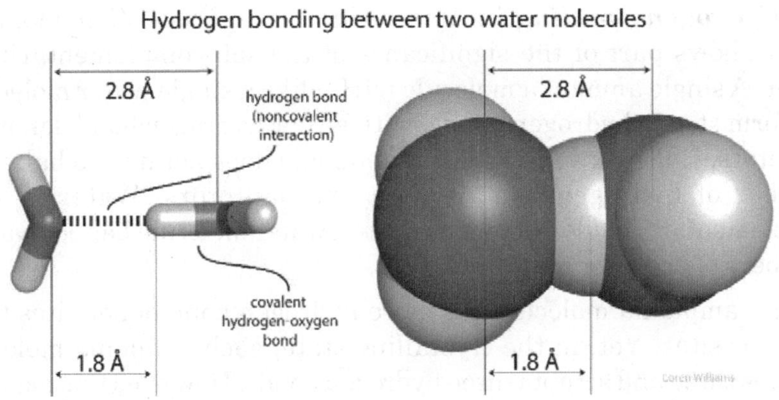

Figure: *Figureshows how hydrogen bonds link two water molecules. This figure illustrates the difference between a covalent bond, linking an oxygen atom to a hydrogen atom, and a hydrogen bond, also linking an oxygen to a hydrogen. A hydrogen bond is a non-covalent molecular interaction. Oxygen atoms are red and hydrogen atoms are white. The space filling representation on the right shows how hydrogen bonding causes violations of van der Waals surfaces.*

Oxygen is highly electronegative, and gains partial negative charge by withdrawing electron density from the two hydrogen atoms to which it is covalently bonded, leaving them with partial positive charges. Water has a balanced number of hydrogen bond donors and acceptors. In ice, every water molecule acts as a donor in two hydrogen bonds and an acceptor in two hydrogen bonds. Water is an excellent hydrogen bonding solvent. The coordinates of a water molecule linked by hydrogen bonds to two other water molecules are here.

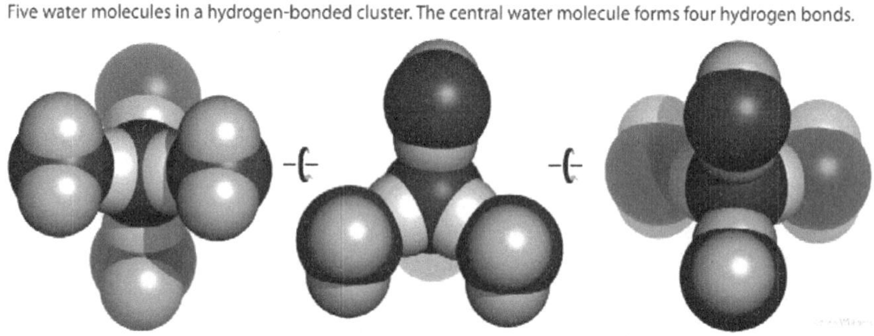

Figure: *Figure illustrates that a water molecule can donate two hydrogen bonds and accept two hydrogen bonds. The central water molecule here is donating two and accepting two hydrogen bonds. In bulk liquid water the total number of hydrogen bond donors equals the total number of hydrogen bond acceptors. All hydrogen bonding donors and acceptors are satisfied. Water is self-complementary.*

Hydrogen Bonding in Ammonia: A comparison of ammonia to water shows part of the significance of the self-complementarity of water. A single ammonia molecule (NH_3), like a single water molecule, can form strong hydrogen bonds, with either hydrogen bond donors or acceptors. However unlike water, ammonia does not have a balanced number of hydrogen bond donors and acceptors. Water is self-complementary, while ammonia is not. A water molecule has equivalent numbers of acceptor and donor sites.

An ammonia molecule has more hydrogen bond donor sites than acceptor sites. Yet, in the crystalline state, each ammonia molecule donates three and accepts three hydrogen bonds. How does this happen? To achieve six hydrogen bonds per ammonia molecule, the single lone pair orbital on each nitrogen is shared by three hydrogen bond donors (N-H's) of three adjacent ammonia molecules. This hydrogen bond is trifucated, as described above. All of these hydrogen bonds are sub-optimal. Each hydrogen bond in crystalline ammonia is long, bent and weak.

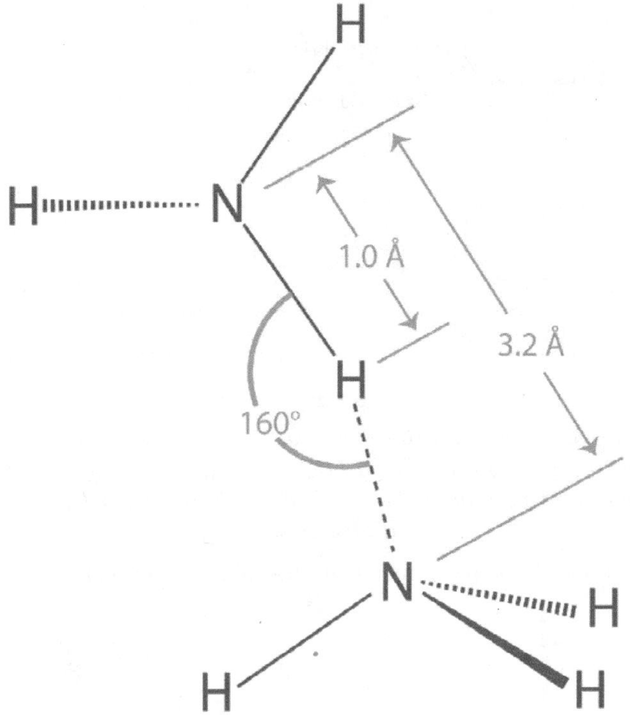

Figure: *Figure Illustrates the hydrogen bonding as observed in crystalline ammonia. The hydrogen bonds are longer than those in ice and are non-linear. Although each ammonia molecule forms hydrogen bonds with six neighbors in the crystal, only two ammonia molecules are shown here.*

The boiling point of ammonia is –33 °C, much lower than that of water (100 °C), indicating that molecular interactions in NH_3 (liq) are significantly weaker than in H_2O (liq).

Although the number of hydrogen bonds per molecule is greater in solid/liquid ammonia than in water, the hydrogen bonds in water are much stronger. The coordinates of an ammonia molecule are here.

Cooperativity of Hydrogen Bonds: In biological systems, hydrogen bonds are frequently cooperative. In systems with multiple hydrogen bonds, the strength of one hydrogen bond is increased by a adjacent hydrogen bond.

For example in the hydrogen-bonded systems below (the acetic acid dimer), the top hydrogen bond increases both the acidity of the hydrogen, and the basicity of the oxygen in the bottom hydrogen bond. Each hydrogen bond makes the other stronger than it would be in isolation. Cooperativity of hydrogen bonding is observed in base pairing and in folded proteins.

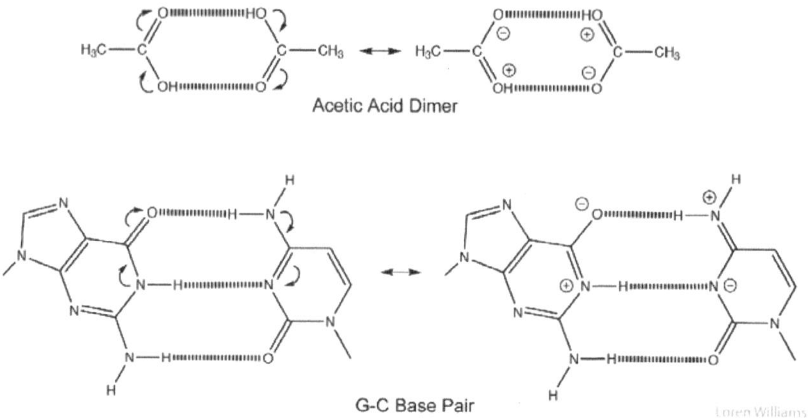

Figure: *Figure shows cooperativity via resonance of the hydrogen bonds of an acetic acid dimer (top) and of a G-C base pair (bottom). Formation of one hydrogen bond increases the stability of an adjacent hydrogen bond (and vice versa).*

Hydrogen bonding cooperativity in antiparallel β-sheets

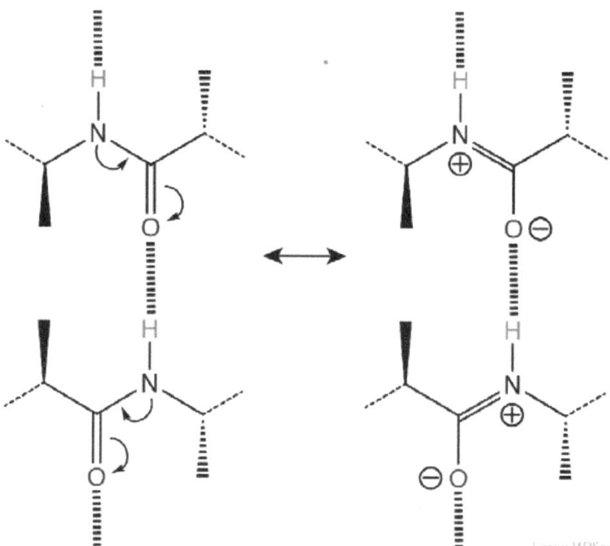

Figure: *Figure shows cooperativity via resonance of the hydrogen bonds of an anti-parallel β-sheet.*

Hydrogen Bonding in Molecular Recognition and Assembly: Because of their directionality, tunability, and ubiquity in simple organic molecules and biological polymers, hydrogen bonding interactions are one of nature's most powerful devices of molecular recognition. Hydrogen bonding donors and acceptors, in complementary 2D and 3D arrays, are observed in many biological assemblies. The locations and directions of the donors and the acceptors are matched, sometimes over vast surfaces. However, not all complementary surfaces in biology involve hydrogen bonds. Leucine zippers, between α-helices, are examples of complementary interactions that involve molecular interactions other than hydrogen bonds.

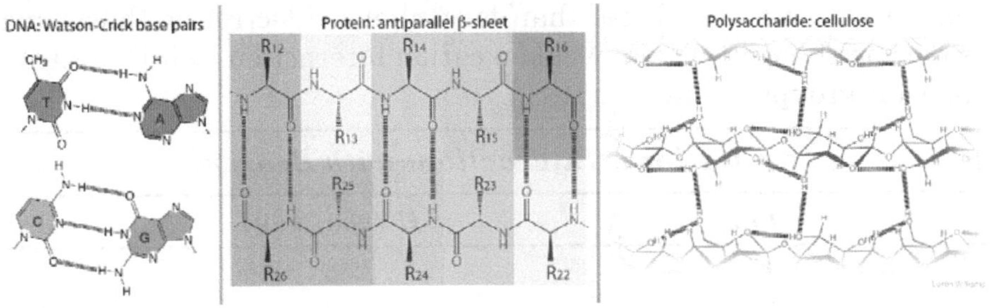

Figure: *Self assembly of biological macromolecules is driven by complementary hydrogen-bonding interactions. (Left) Base pairing between complementary hydrogen bond donors and acceptors on the sidechains of nucleic acids. (Center) Backbone assembly between self-complementary hydrogen bond donors and acceptors of the protein backbone to form anti-parallel β-strands in a β-sheet, and (Right) Self-complementary hydrogen bond donors and acceptors in carbohydrate, between glucose moieties within cellulose.*

Hydrogen Bonding in Template-Directed Catalysis: Biological systems have unique abilities to link complex molecular interactions to catalytic functions. Sophisticated non-covalent interactions control formation of covalent bonds. Some of the most advanced forms of these phenomena are observed in DNA and RNA polymerases, and in the ribosome. In these systems hydrogen bonding and other molecular interactions direct catalytic function. In an RNA polymerase, if 'correct' hydrogen bonding (i.e., C-G or A-U/T Watson-Crick hydrogen bonding) occurs between the template strand and the incoming nucleotide, then the enzyme catalyzes formation of a covalent bond. When 'wrong' interactions (e.g., a G-U pair) are detected, the enzyme kicks out the incoming nucleotide without forming the covalent bond. Therefore one molecule acts as a template that directs synthesis of another molecule, in close analogy with the way that a pastry template directs the shape of the pastry.

Cation-π Interactions

A π-system such as benzene, tryptophan, phenylalanine or tyrosine focuses partial negative charge above and below the plane of the ring. A cation can interact favourably with this negative charge when the cation is near the face of the π-system. In the most stable complexes of this type, the cation is centred directly over the π-system and is in direct van der Waals contact with it. The table on the left shows gas phase interaction enthalpies, which are on the same order as the hydration enthalpies for these cations. Therefore, cation-π interactions are roughly similar in strength to cation-dipole interactions formed between water and cations. Small ions with high charge density form stonger cation-π complexes than larger ions. Electron withdrawing groups on the ring system weaken cation interactions while electron donating groups strengthen them.

Cation-π Interactions with Benzene

Ion	ΔH *(kcal/mol)*
Li^{+}	-38
Na^{+}	-28
K^{+}	-19
NH_4^{+}	-19

(gas phase)

Cation-π interactions are important in protein structure. The guanidinium group of arginine and the ε-NH_3^{+} of lysine engage in cation-π interactions with aromatic protein sidechains. A favourable cation-π pair contributes as much to protein stability as a good hydrogen bond or an electrostatic (charge-charge) interaction. Tryptophan is the most frequent π system in protein cation-π pairs while arginine is the most frequent cation. Tryptophan and arginine can form extended coplaner assemblies.

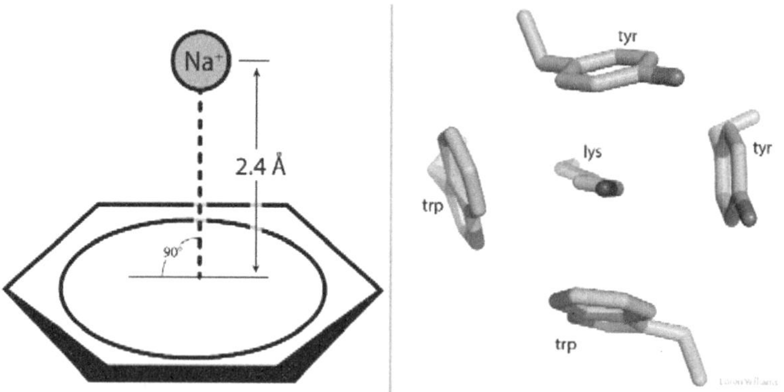

Figure: *(Left) Optimal geometry for a cation-π interaction between a Na+ cation and a benzene molecule. The distance from the Na+ to the center of the ring is 2.4 Å (ionic radius of Na+ = 0.9 Å, vdw radius of C (rC) = 1.7 Å). (Right) The ε-NH3+ of a lysine engages in cation-π interactions with two tryptophan sidechains and two tyrosine sidechains (glucoamylase, PDB ID 1GAI).*

The Hydrophobic Effect

Water and Biology: Biology and biochemistry take place in complex aqueous environments. Living organisms are around 80% water by weight. Life as we know it on earth is fully dependent on and intertwined with water. Water is a crucial determinant of structures and properties of cellular assemblies and organelles and of biochemical reactions. Water is a reactive chemical and is a direct and critical participant in some of the most central and universal reactions of biology (hydrolysis and dehydration/condensation). All of life's polymers are synthesized by dehydration/condensation and all are broken down by hydrolysis.

Water is a powerful solvent for ions and polar substances and is a poor solvent for non-polar substances. Water causes certain amphipathic molecules (with both polar and non-polar functionalities) to spontaneously form compartments. In water, membranes assemble and proteins fold.

The unusual cohesion of water molecules can be inferred from water's high melting point, boiling point, heat of vaporization, heat of fusion and surface tension and by water's increase in volume upon freezing. Each of these parameters indicates that water is a special liquid. For example the heat of vaporization of water (540 cal/g) is over twice that of methanol (263 cal/g) and nearly ten times that of chloroform (59 cal/g).

Water has a unique ability to shield charged species from each other. Electrostatic interactions between ions are highly attenuated in water. The electrostatic force between two ions in solution is inversely proportional to the dielectric constant of the solvent. The dielectric constant of water (80.0) is huge. It is over twice that of methanol (33.1) and over five times that of ammonia (15.5). Water is a good solvent for salts because the attractive forces between cations and anions are minimized by water.

The hydrophobic effect can be understood only after thinking carefully about water. The hydrophobic effect is a consequence of strong directional interactions between water molecules and the complementarity of those interactions.

Molecular Structure of Water in the Crystalline State: A water molecule has four filled valence orbitals (sp^3 hybridized) that form a modestly distorted tetrahedron. Two of the electron pairs form covalent bonds with hydrogen atoms and two are non-bonding. The non-bonding lone pairs take more space than the bonding lone pairs, causing the distortion from a perfect tetrahedron. It is useful to imagine that a water molecule is a tetrahedron with negative charge on two apexes and positive charge on two apexes. Ideally, every water molecule can interact with exactly four other water molecules.

Oxygen, which is highly electronegative, withdraws electron density from the hydrogen atoms to the extent that they are essentially bare protons on their exposed sides (distal to the oxygen). The charge distribution of a water molecule (partial negative charge on oxygen and partial positive charge on hydrogen) is shown below.

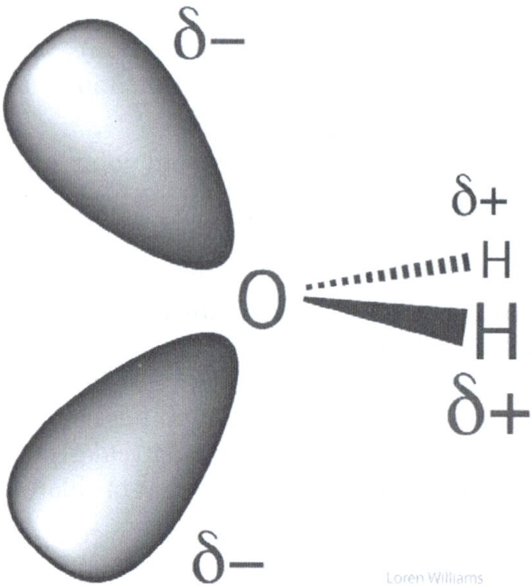

Figure: *Figure illustrates the two lone electron pairs and the two bonding electron pairs of a water molecule. A four valence orbitals of a water molecule form a slightly distorted tetrahedron. The non-bonding electron pairs take up a little more space than the bonding electron pairs.*

X-ray and neutron diffraction of crystalline ice shows that each water molecule is engaged in four hydrogen bonds with intermolecular oxygen-oxygen distances of 2.76 Å. Each oxygen atom is located at the centre of a tetrahedron formed by four other oxygen atoms. Each hydrogen atom lies on a line between two oxygen atoms and forms a covalent bond to one oxygen (bond length: 1.00 Å) and a hydrogen bond to the other (hydrogen bond length: 1.76 Å). The tetrahedral shape of an individual water molecule is projected out into the surrounding crystal lattice. The hydrogen atoms are not located midway between oxygen atoms.

Water molecules in the crystalline state are not closely packed, resulting in tiny cavities of empty space within the crystal. The cavities are formed because the directionality of water-water interactions dominates water-water packing considerations. Small cavities in the solid lattice but not in the liquid are the reason that water increases in volume upon freezing (i.e., ice floats). There are many degrees of freedom in hydrogen bond donor/acceptor relationships that are interconverted by cooperative rotations. Ice is rather disordered in that respect.

Molecular Structure of Water in the Liquid State: In the liquid state, water is not as ordered as in the crystalline state. In the liquid state at O degrees C a time-averaged water molecule is involved in around 3.5 intermolecular hydrogen bonds. Some of them are three- and four-centred. Liquid water is more dense than solid water. Nevertheless, the macroscopic properties of liquid water are dominated by the directional and complementary cohesive interactions between water molecules.

Empirical Description of the Hydrophobic Effect: Water and oil do not mix. Non-polar substances are not soluble in water. That is the empirical description of hydrophobic effect. Hydrophobic substances are those that are soluble in non-polar solvents (such as CCl_4 or cyclohexane) but are only slightly soluble in water. Hydrophobic substances are non-polar. The definition excludes substances like cellulose which are generally insoluble because of strong intermolecular cohesion. Hydrophobic substances are structurally incapable of forming hydrogen bonds. Hydrocarbons ($CH_3CH_2CH_2 \cdots CH_2CH_3$) are hydrophobic.

Molecular Basis of the Hydrophobic Effect: Understanding the molecular and thermodynamic nature of the hydrophobic effect is not easy. Many textbooks contain superficial or incorrect explanations. The most important thing to remember is that the hydrophobic effect is fully a function of water, it a consequence of the distinctive molecular structure and self-assembly properties of water. Hydrophobic substances are essentially passive participants in hydrophobic processes such as the separation of oil from water. The molecular interactions of a hydrocarbon with neighbouring water molecules in aqueous solution are just as favourable as with neighbouring hydrocarbon molecules in pure liquid hydrocarbon. A hydrocarbon molecule is just as happy (forms equally favourable molecular interactions) in aqueous solution as in neat (pure) hydrocarbon.

The molecular interactions of a water molecule adjacent to a hydrocarbon (or other non-polar molecule) are just as strong and favourable (in terms of enthalpy) as the interactions of a water molecule in bulk water, surrounded by water only. There is no net change in favourable molecular interactions when oil and water mix or separate.

If it is not because of changes in molecular interactions, why don't oil and water mix? Why do they spontaneously separate? The reason is that water molecules adjacent a hydrocarbon sacrifice rotational and translational freedom to maintain molecular interactions. Water adjacent to a hydrocarbon pays the price of low entropy to maintain good molecular interactions. The strong directional cohesive interactions between water molecules are maintained, but at a high entropic cost. The low entropy of the water in the interfacial region (ie directly adjacent to a hydrocarbon molecule) arises from the strong directional forces between water molecules. In bulk water, these forces are essentially isotropic (extending in all directions). At the interface these forces are anisotropic because the cyclohexane molecule does not form hydrogen bonds. So an entropic effect leads to an unfavourable free energy of mixing oil and water ($\Delta G = \Delta H - T\Delta S > 0$)

The term 'hydrophobic bond' is a misnomer and should be avoided, even though Walter Kauzmann, the discoverer of the hydrophobic effect, did often use that phrase.

Thermodynamic Basis of the Hydrophobic Effect: The molecular descriptions of the hydrophobic effect above can be understood by the thermodynamic parameters enthalpy (ΔH, indicates changes in molecular interactions) and entropy (ΔS, indicates changes in available rotational, translational, vibrational states, etc).

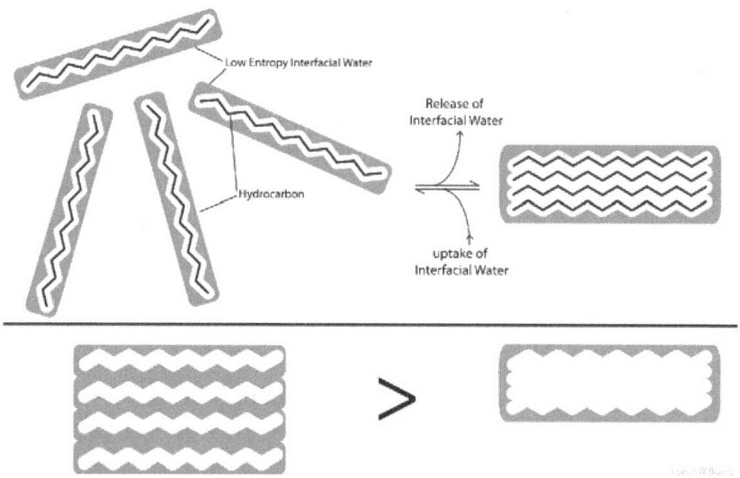

Figure: *Figure shows how aggregation of hydrocarbon molecules causes the release of interfacial water molecules. Therefore the system gains entropy (positive T?S) upon hydrocarbon aggregation. Release of low entropy interfacial water molecules into the bulk solution drives hydrocarbon aggregation. The bottom panel illustrates that there is more interfacial water on the left hand side of the equation than on the right hand side.*

A hydrocarbon engages in favourable molecule interactions with water in aqueous solution. We know this because the transfer of a mole of hydrocarbon from pure hydrocarbon to dilute aqueous solution has an enthalpy of around zero. So why don't oil and water mix? It is the water. Water drives non-polar substances out of the aqueous phase.

As illustrated above, in the aqueous phase a region of relatively low entropy (high order) water forms at the interface between the aqueous solvent and a hydrophobic solute.

When isolated hydrocarbon molecules aggregate in aqueous solution, the total volume of interfacial water decreases. Thus the driving force for aggregation of hydrophobic substances arises from an increase in entropy of the water. The driving force for aggregation does not arise from intrinsic attraction between hydrophobic solute molecules.

If one considers the entropy of the hydrocarbon molecules alone, a dispersed solution has greater entropy, and is more stable, than an aggregated state. Similarly, a protein may appear to have greater entropy in a random coil than in a native state. Only when the entropy of the aqueous phase is factored into the equation can one understand the separation of water and oil into two phases, and the folding of a protein into a native state.

Counterion Release

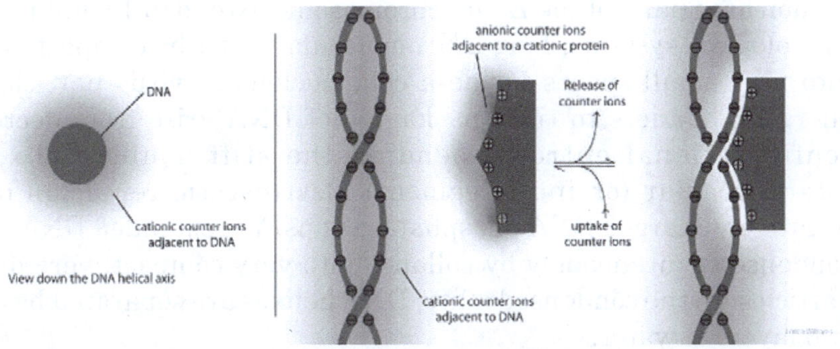

Figure: *(left) Figure shows an axial view of DNA, represented as a anionic cyclinder. Cationic counterions (orange shading) surround the cyclinder. The concentration of cations decreases with distance from the surface of the cyclinder. The deeper orange shading indicates more concentrated cations. The panel on the right illustrates how both anionic counterions (blue) associated with a cationic protein, and cationic counterions (orange) associated with anionic DNA, are released to bulk solution when the protein binds to DNA. This release of counterions drives the association (by contributing +T?S).*

For many purposes it is useful to think of DNA as a rod that is coated with anionic charge. In aqueous solution the negative rod is surrounded by cations such as Na^+, K^+ and Mg^{2+} and/or by polyamines. The high density of negative charge on the rod causes strong radial electric fields. These electric fields lead to steep radial gradients of the cation concentrations.

Theoretical considerations (counterion condensation) predict that the local concentration of a monovalent cation such as K^+ near the surface of DNA is around 2 Molar. It is counter intuitive, but the concentration of K^+ surrounding DNA is largely independent of the K^+ concentration in bulk solution. The electrostatic environment surrounding DNA does not depend on the bulk concentration of salt.

Counterions are released when a cationic protein binds to DNA. This release explains the dramatic salt dependencies of DNA-protein complexes. High salt destabilizes DNA-protein complexes. If the bulk salt concentration is low, there is a large entropic gain from counterion release, and the protein binds tightly to the DNA. If the bulk salt concentration is high, the entropic gain from counterion release is small, and the protein binds weakly. Counter ion release explains much of the salt dependencies of DNA melting, RNA folding and DNA condensation.

DNA condensation. Genomic DNAs are very long molecules. The 160,000 base pairs of T4 phage DNA extend to 54 microns. The 4.2 million base pairs of the *E. coli* chromosome extend to 1.4 millimetres. In biological systems, long DNA molecules must be compacted to fit into very small spaces inside a cell, nucleus or virus particle. The energetic barriers to tight packaging of DNA arise from decreased configurational entropy, bending the stiff double helix, and intermolecular (or inter-segment) electrostatic repulsion of the negatively charged DNA phosphate groups. Yet extended DNA chains condense spontaneously by collapse into very compact, very orderly particles. In the condensed state, DNA helixes are separated by one or two layers of water.

Condensed DNA particles are commonly compact toroids. DNA condensation in aqueous solution requires highly charged cations such as spermine (+4) or spermidine (+3). Divalent cations will condense DNA in water-alcohol mixtures. The role of the cations is to decrease electrostatic repulsion of adjacent negatively charged DNA segments. The source of the attraction between nearby DNA segments is not so easy to understand. One possible source of attraction are fluctuations of ion atmospheres in analogy with fluctuating dipoles between molecules

Metabolism

Metabolism is the set of life-sustaining chemical transformations within the cells of living organisms. These enzyme-catalyzed reactions allow organisms to grow and reproduce, maintain their structures, and respond to their environments. The word metabolism can also refer to all chemical reactions that occur in living organisms, including digestion and the transport of substances into and between different cells, in which case the set of reactions within the cells is called intermediary metabolism or intermediate metabolism.

Metabolism is usually divided into two categories. Catabolism, that breaks down organic matter and harvests energy by way of cellular respiration, and anabolism that uses energy to construct components of cells such as proteins and nucleic acids.

The chemical reactions of metabolism are organized into metabolic pathways, in which one chemical is transformed through a series of steps into another chemical, by a sequence of enzymes. Enzymes are crucial to metabolism because they allow organisms to drive desirable reactions that require energy that will not occur by themselves, by coupling them to spontaneous reactions that release energy. Enzymes act as catalysts that allow the reactions to proceed more rapidly. Enzymes also allow the regulation of metabolic pathways in response to changes in the cell's environment or to signals from other cells.

The metabolic system of a particular organism determines which substances it will find nutritious and which poisonous. For example, some prokaryotes use hydrogen sulfide as a nutrient, yet this gas is poisonous to animals. The speed of metabolism, the metabolic rate, influences how much food an organism will require, and also affects how it is able to obtain that food.

A striking feature of metabolism is the similarity of the basic metabolic pathways and components between even vastly different species. For example, the set of carboxylic acids that are best known as the intermediates in the citric acid cycle are present in all known organisms, being found in species as diverse as the unicellular bacterium *Escherichia coli* and huge multicellular organisms like elephants. These striking similarities in metabolic pathways are likely due to their early appearance in evolutionary history, and retained because of their efficacy.

Key Biochemicals

Most of the structures that make up animals, plants and microbes are made from three basic classes of molecule: amino acids, carbohydrates and lipids (often called fats).

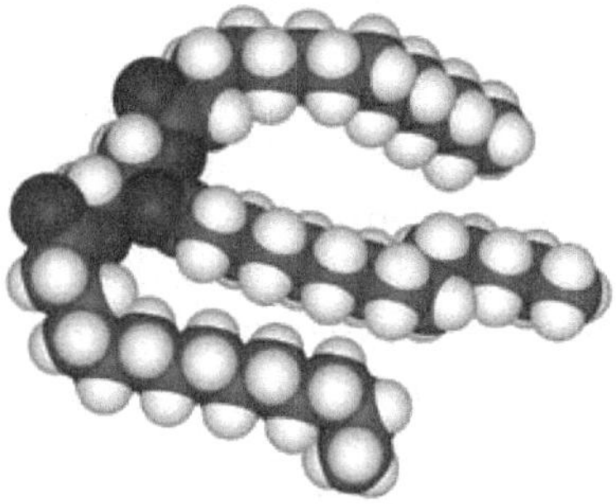

***Figure:** Structure of a triacylglycerol lipid*

As these molecules are vital for life, metabolic reactions either focus on making these molecules during the construction of cells and tissues, or by breaking them down and using them as a source of energy, by their digestion. These biochemicals can be joined together to make polymers such as DNA and proteins, essential macromolecules of life.

Type of molecule	*Name of monomer forms*	*Name of polymer forms*	*Examples of polymer forms*
Amino acids	Amino acids	Proteins (also called polypeptides)	Fibrous proteins and globular proteins
Carbohydrates	Monosaccharides	Polysaccharides	Starch, glycogen and cellulose
Nucleic acids	Nucleotides	Polynucleotides	DNA and RNA

Amino Acids and Proteins

Proteins are made of amino acids arranged in a linear chain joined together by peptide bonds. Many proteins are enzymes that catalyze the chemical reactions in metabolism. Other proteins have structural or mechanical functions, such as those that form the cytoskeleton, a system of scaffolding that maintains the cell shape. Proteins are also important in cell signaling, immune responses, cell adhesion, active transport across membranes, and the cell cycle. Amino acids also contribute to cellular energy metabolism by providing a carbon source for entry into the citric acid cycle (tricarboxylic acid cycle), especially when a primary source of energy, such as glucose, is scarce, or when cells undergo metabolic stress.

Lipids

Lipids are the most diverse group of biochemicals. Their main structural uses are as part of biological membranes both internal and external, such as the cell membrane, or as a source of energy. Lipids are usually defined as hydrophobic or amphipathic biological molecules but will dissolve in organic solvents such as benzene or chloroform. The fats are a large group of compounds that contain fatty acids and glycerol; a glycerol molecule attached to three fatty acid esters is called a triacylglyceride. Several variations on this basic structure exist, including alternate backbones such as sphingosine in the sphingolipids, and hydrophilic groups such as phosphate as in phospholipids. Steroids such as cholesterol are another major class of lipids.

Carbohydrates

Carbohydrates are aldehydes or ketones, with many hydroxyl groups attached, that can exist as straight chains or rings. Carbohydrates are the most abundant biological molecules, and fill numerous roles, such as the storage and transport of energy (starch, glycogen) and structural components (cellulose in plants, chitin in animals). The basic carbohydrate units are called monosaccharides and include galactose, fructose, and most importantly glucose. Monosaccharides can be linked together to form polysaccharides in almost limitless ways.

Nucleotides

The two nucleic acids, DNA and RNA are polymers of nucleotides, each nucleotide is composed of a phosphate group made up of a ribose sugar group with a nitrogenous base. Nucleic acids are critical for the storage and use of genetic information, and its interpretation through the processes of transcription and protein biosynthesis. This information is protected by DNA repair mechanisms and propagated through DNA replication. Many viruses have an RNA genome, for example HIV, which uses reverse transcription to create a DNA template from its viral RNA genome. RNA in ribozymes such as spliceosomes and ribosomes is similar to enzymes as it can catalyze chemical reactions. Individual nucleosides are made by attaching a nucleobase to a ribose sugar. These bases are heterocyclic rings containing nitrogen, classified as purines or pyrimidines. Nucleotides also act as coenzymes in metabolic-group-transfer reactions.

Coenzymes

Metabolism involves a vast array of chemical reactions, but most fall under a few basic types of reactions that involve the transfer of functional groups of atoms and their bonds within molecules. This common chemistry allows cells to use a small set of metabolic intermediates to carry chemical groups between different reactions. These group-transfer intermediates are called coenzymes. Each class of group-transfer reactions is carried out by a particular coenzyme, which is the substrate for a set of enzymes that produce it, and a set of enzymes that consume it. These coenzymes are therefore continuously made, consumed and then recycled.

One central coenzyme is adenosine triphosphate (ATP), the universal energy currency of cells. This nucleotide is used to transfer chemical energy between different chemical reactions. There is only a small amount of ATP in cells, but as it is continuously regenerated, the human body can use about its own weight in ATP per day. ATP acts as a bridge between catabolism and anabolism. Catabolism breaks down molecules and anabolism puts them together. Catabolic reactions generate ATP and anabolic reactions consume it. It also serves as a carrier of phosphate groups in phosphorylation reactions.

A vitamin is an organic compound needed in small quantities that cannot be made in cells. In human nutrition, most vitamins function as coenzymes after modification; for example, all water-soluble vitamins are phosphorylated or are coupled to nucleotides when they are used in cells. Nicotinamide adenine dinucleotide (NAD^+), a derivative of vitamin B_3 (niacin), is an important coenzyme that acts as a hydrogen acceptor. Hundreds of separate types of dehydrogenases remove electrons from their substrates and reduce NAD^+ into NADH. This reduced form of the coenzyme is then a substrate for any of the reductases in the cell that need to reduce their substrates. Nicotinamide adenine dinucleotide exists in two related forms in the cell, NADH and NADPH. The NAD^+/NADH form is more important in catabolic reactions, while $NADP^+$/NADPH is used in anabolic reactions.

Minerals and Cofactors

Inorganic elements play critical roles in metabolism; some are abundant (e.g. sodium and potassium) while others function at minute concentrations. About 99% of a mammal's mass is made up of the elements carbon, nitrogen, calcium, sodium, chlorine, potassium, hydrogen, phosphorus, oxygen and sulfur. Organic compounds (proteins, lipids and carbohydrates) contain the majority of the carbon and nitrogen; most of the oxygen and hydrogen is present as water.

The abundant inorganic elements act as ionic electrolytes. The most important ions are sodium, potassium, calcium, magnesium, chloride, phosphate and the organic ion bicarbonate. The maintenance of precise ion gradients across cell membranes maintains osmotic pressure and pH. Ions are also critical for nerve and muscle function, as action potentials in these tissues are produced by the exchange of electrolytes between the extracellular fluid and the cell's fluid, the cytosol. Electrolytes enter and leave cells through proteins in the cell membrane called ion channels. For example, muscle contraction depends upon the movement of calcium, sodium and potassium through ion channels in the cell membrane and T-tubules.

Transition metals are usually present as trace elements in organisms, with zinc and iron being most abundant of those. These metals are used in some proteins as cofactors and are essential for the activity of enzymes such as catalase and oxygen-carrier proteins such as hemoglobin. Metal cofactors are bound tightly to specific sites in proteins; although enzyme cofactors can be modified during catalysis, they always return to their original state by the end of the reaction catalyzed. Metal micronutrients are taken up into organisms by specific transporters and bind to storage proteins such as ferritin or metallothionein when not in use.

Catabolism

Catabolism is the set of metabolic processes that break down large molecules. These include breaking down and oxidizing food molecules. The purpose of the catabolic reactions is to provide the energy and components needed by anabolic reactions. The exact nature of these catabolic reactions differ from organism to organism and organisms can be classified based on their sources of energy and carbon (their primary nutritional groups), as shown in the table below. Organic molecules are used as a source of energy by organotrophs, while lithotrophs use inorganic substrates and phototrophs capture sunlight as chemical energy. However, all these different forms of metabolism depend on redox reactions that involve the transfer of electrons from reduced donor molecules such as organic molecules, water, ammonia, hydrogen sulfide or ferrous ions to acceptor molecules such as oxygen, nitrate or sulfate. In animals these reactions involve complex organic molecules that are broken down to simpler molecules, such as carbon dioxide and water. In photosynthetic organisms such as plants and cyanobacteria, these electron-transfer reactions do not release energy, but are used as a way of storing energy absorbed from sunlight.

<table>
<tr><th colspan="6">Classification of organisms based on their metabolism</th></tr>
<tr><td rowspan="2">Energy source</td><td>sunlight</td><td>photo-</td><td rowspan="2" colspan="2"></td><td rowspan="6">-
troph</td></tr>
<tr><td>Preformed molecules</td><td>chemo-</td></tr>
<tr><td rowspan="2">Electron donor</td><td>organic compound</td><td rowspan="2"></td><td>organo-</td><td rowspan="2"></td></tr>
<tr><td>inorganic compound</td><td>litho-</td></tr>
<tr><td rowspan="2">Carbon source</td><td>organic compound</td><td rowspan="2" colspan="2"></td><td>hetero-</td></tr>
<tr><td>inorganic compound</td><td>auto-</td></tr>
</table>

The most common set of catabolic reactions in animals can be separated into three main stages. In the first, large organic molecules such as proteins, polysaccharides or lipids are digested into their smaller components outside cells. Next, these smaller molecules are taken up by cells and converted to yet smaller molecules, usually acetyl coenzyme A (acetyl-CoA), which releases some energy. Finally, the acetyl group on the CoA is oxidised to water and carbon dioxide in the citric acid cycle and electron transport chain, releasing the energy that is stored by reducing the coenzyme nicotinamide adenine dinucleotide (NAD^+) into NADH.

Digestion

Macromolecules such as starch, cellulose or proteins cannot be rapidly taken up by cells and must be broken into their smaller units before they can be used in cell metabolism. Several common classes of enzymes digest these polymers. These digestive enzymes include proteases that digest proteins into amino acids, as well as glycoside hydrolases that digest polysaccharides into simple sugars known as monosaccharides.

Microbes simply secrete digestive enzymes into their surroundings, while animals only secrete these enzymes from specialized cells in their guts. The amino acids or sugars released by these extracellular enzymes are then pumped into cells by active transport proteins.

Energy from Organic Compounds

Carbohydrate catabolism is the breakdown of carbohydrates into smaller units. Carbohydrates are usually taken into cells once they have been digested into monosaccharides. Once inside, the major route of breakdown is glycolysis, where sugars such as glucose and fructose are converted into pyruvate and some ATP is generated. Pyruvate is an intermediate in several metabolic pathways, but the majority is converted to acetyl-CoA and fed into the citric acid cycle. Although some more ATP is generated in the citric acid cycle, the most important product is NADH, which is made from NAD^+ as the acetyl-CoA is oxidized. This oxidation releases carbon dioxide as a waste product. In anaerobic conditions, glycolysis produces lactate, through the enzyme lactate dehydrogenase re-oxidizing NADH to NAD+ for re-use in glycolysis. An alternative route for glucose breakdown is the pentose phosphate pathway, which reduces the coenzyme NADPH and produces pentose sugars such as ribose, the sugar component of nucleic acids.

Fats are catabolised by hydrolysis to free fatty acids and glycerol. The glycerol enters glycolysis and the fatty acids are broken down by beta oxidation to release acetyl-CoA, which then is fed into the citric acid cycle. Fatty acids release more energy upon oxidation than carbohydrates because carbohydrates contain more oxygen in their structures. Steroids are also broken down by some bacteria in a process similar to beta oxidation, and this breakdown process involves the release of significant amounts of acetyl-CoA, propionyl-CoA, and pyruvate, which can all be used by the cell for energy. *M. tuberculosis* can also grow on the lipid cholesterol as a sole source of carbon, and genes involved in the cholesterol use pathway(s) have been validated as important during various stages of the infection lifecycle of *M. tuberculosis*.

Amino acids are either used to synthesize proteins and other biomolecules, or oxidized to urea and carbon dioxide as a source of energy. The oxidation pathway starts with the removal of the amino group by a transaminase. The amino group is fed into the urea cycle, leaving a deaminated carbon skeleton in the form of a keto acid. Several of these keto acids are intermediates in the citric acid cycle, for example the deamination of glutamate forms α-ketoglutarate. The glucogenic amino acids can also be converted into glucose, through gluconeogenesis.

Pumping protons out of the mitochondria creates a proton concentration difference across the membrane and generates an electrochemical gradient. This force drives protons back into the mitochondrion through the base of an enzyme called ATP synthase. The flow of protons makes the stalk subunit rotate, causing the active site of the synthase domain to change shape and phosphorylate adenosine diphosphate – turning it into ATP.

Energy from Inorganic Compounds

Chemolithotrophy is a type of metabolism found in prokaryotes where energy is obtained from the oxidation of inorganic compounds. These organisms can use hydrogen, reduced sulfur compounds (such as sulfide, hydrogen sulfide and thiosulfate), ferrous iron (FeII) or ammonia as sources of reducing power and they gain energy from the oxidation of these compounds with electron acceptors such as oxygen or nitrite. These microbial processes are important in global biogeochemical cycles such as acetogenesis, nitrification and denitrification and are critical for soil fertility.

Energy from Light

The energy in sunlight is captured by plants, cyanobacteria, purple bacteria, green sulfur bacteria and some protists. This process is often coupled to the conversion of carbon dioxide into organic compounds, as part of photosynthesis, which is discussed below. The energy capture and carbon fixation systems can however operate separately in prokaryotes, as purple bacteria and green sulfur bacteria can use sunlight as a source of energy, while switching between carbon fixation and the fermentation of organic compounds.

In many organisms the capture of solar energy is similar in principle to oxidative phosphorylation, as it involves the storage of energy as a proton concentration gradient. This proton motive force then drives ATP synthesis. The electrons needed to drive this electron transport chain come from light-gathering proteins called photosynthetic reaction centres or rhodopsins. Reaction centres are classed into two types depending on the type of photosynthetic pigment present, with most photosynthetic bacteria only having one type, while plants and cyanobacteria have two.

In plants, algae, and cyanobacteria, photosystem II uses light energy to remove electrons from water, releasing oxygen as a waste product. The electrons then flow to the cytochrome b6f complex, which uses their energy to pump protons across the thylakoid membrane in the chloroplast. These protons move back through the membrane as they drive the ATP synthase, as before. The electrons then flow through photosystem I and can then either be used to reduce the coenzyme $NADP^+$, for use in the Calvin cycle, which is discussed below, or recycled for further ATP generation.

Anabolism

Anabolism is the set of constructive metabolic processes where the energy released by catabolism is used to synthesize complex molecules. In general, the complex molecules that make up cellular structures are constructed step-by-step from small and simple precursors. Anabolism involves three basic stages. Firstly, the production of precursors such as amino acids, monosaccharides, isoprenoids and nucleotides, secondly, their activation into reactive forms using energy from ATP, and thirdly, the assembly of these precursors into complex molecules such as proteins, polysaccharides, lipids and nucleic acids.

Organisms differ in how many of the molecules in their cells they can construct for themselves. Autotrophs such as plants can construct the complex organic molecules in cells such as polysaccharides and proteins from simple molecules like carbon dioxide and water. Heterotrophs, on the other hand, require a source of more complex substances, such as monosaccharides and amino acids, to produce these complex molecules. Organisms can be further classified by ultimate source of their energy: photoautotrophs and photoheterotrophs obtain energy from light, whereas chemoautotrophs and chemoheterotrophs obtain energy from inorganic oxidation reactions.

Carbon Fixation

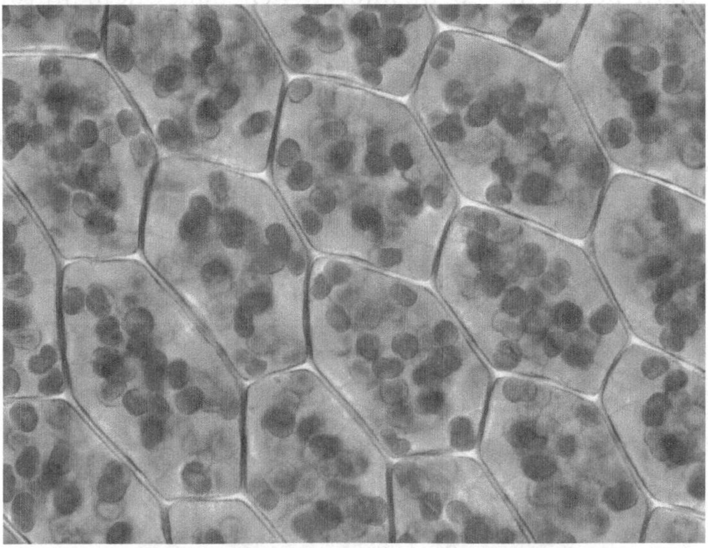

Figure: *Plant cells (bounded by purple walls) filled with chloroplasts (green), which are the site of photosynthesis*

Photosynthesis is the synthesis of carbohydrates from sunlight and carbon dioxide (CO_2). In plants, cyanobacteria and algae, oxygenic photosynthesis splits water, with oxygen produced as a waste product. This process uses the ATP and NADPH produced by the photosynthetic reaction centres, as described above, to convert CO_2 into glycerate 3-phosphate, which can then be converted into glucose. This carbon-fixation reaction is carried out by the enzyme RuBisCO as part of the Calvin – Benson cycle. Three types of photosynthesis occur in plants, C3 carbon fixation, C4 carbon fixation and CAM photosynthesis. These differ by the route that carbon dioxide takes to the Calvin cycle, with C3 plants fixing CO_2 directly, while C4 and CAM photosynthesis incorporate the CO_2 into other compounds first, as adaptations to deal with intense sunlight and dry conditions.

In photosynthetic prokaryotes the mechanisms of carbon fixation are more diverse. Here, carbon dioxide can be fixed by the Calvin – Benson cycle, a reversed citric acid cycle, or the carboxylation of acetyl-CoA. Prokaryotic chemoautotrophs also fix CO_2 through the Calvin – Benson cycle, but use energy from inorganic compounds to drive the reaction.Chemosynthesis

In biochemistry, chemosynthesis is the biological conversion of one or more carbon molecules (usually carbon dioxide or methane) and nutrients into organic matter using the oxidation of inorganic molecules (e.g. hydrogen gas, hydrogen sulfide) or methane as a source of energy, rather than sunlight, as in photosynthesis. Chemoautotrophs, organisms that obtain carbon through chemosynthesis, are phylogenetically diverse, but groups that include conspicuous or biogeochemically-important taxa include the sulfur-oxidizing gamma and epsilon proteobacteria, the aquificaeles, the methanogenic archaea and the neutrophilic iron-oxidizing bacteria.

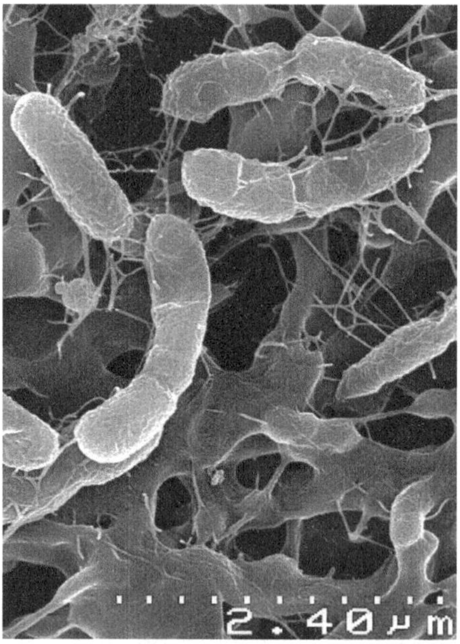

Figure: *Venenivibrio stagnispumantis gains energy by oxidizing hydrogen gas.*

Many microorganisms in dark regions of the oceans also use chemosynthesis to produce biomass from single carbon molecules. Two categories can be distinguished. In the rare sites at which hydrogen molecules (H_2) are available, the energy available from the reaction between CO_2 and H_2 (leading to production of methane, CH_4) can be large enough to drive the production of biomass. Alternatively, in most oceanic environments, energy for chemosynthesis derives from reactions in which substances such as hydrogen sulfide or ammonia are oxidized. This may occur with or without the presence of oxygen.

Many chemosynthetic microorganisms are consumed by other organisms in the ocean, and symbiotic associations between chemosynthesizers and respiring heterotrophs are quite common. Large populations of animals can be supported by chemosynthetic secondary production at hydrothermal vents, methane clathrates, cold seeps, whale falls, and isolated cave water.

It has been hypothesized that chemosynthesis may support life below the surface of Mars, Jupiter's moon Europa, and other planets.

The term chemosynthesis is also used in molecular nanotechnology broadly to refer to any chemical synthesis where reactions occur due to random thermal motion, a class which encompasses almost all of modern synthetic chemistry. The human-authored processes of chemical engineering are accordingly represented as biomimicry of the natural phenomena above, and the entire class of non-photosynthetic chains by which complex molecules are constructed is described as chemo-.

This form of engineering is then contrasted with mechanosynthesis, a hypothetical process where individual molecules are mechanically manipulated to control reactions to human specification. Since photosynthesis and other natural processes create extremely complex molecules to the specifications contained in RNA and stored long-term in DNA form, advocates of molecular engineering claim that an artificial process can likewise exploit a chain of long-term storage, short-term storage, enzyme-like copying mechanisms similar to those in the cell, and ultimately produce complex molecules which need not be proteins. For instance, sheet diamond or carbon nanotubes could be produced by a chain of non-biological reactions that have been designed using the basic model of biology.

Use of the term chemosynthesis reinforces the view that this is feasible by pointing out that several alternate means of creating complex proteins, mineral shells of mollusks and crustaceans, etc., evolved naturally, not all of them dependent on photosynthesis and a food chain from the sun via chlorophyll. Since more than one such pathway exists to creating complex molecules, even extremely specific ones such as proteins edible to fish, the likelihood of humans being able to design an entirely new one is considered (by these advocates) to be near certainty in the long run, and possible within a generation.

Carbohydrates and Glycans

In carbohydrate anabolism, simple organic acids can be converted into monosaccharides such as glucose and then used to assemble polysaccharides such as starch. The generation of glucose from compounds like pyruvate, lactate, glycerol, glycerate 3-phosphate and amino acids is called gluconeogenesis. Gluconeogenesis converts pyruvate to glucose-6-phosphate through a series of intermediates, many of which are shared with glycolysis. However, this pathway is not simply glycolysis run in reverse, as several steps are catalyzed by non-glycolytic enzymes. This is important as it allows the formation and breakdown of glucose to be regulated separately, and prevents both pathways from running simultaneously in a futile cycle.

Although fat is a common way of storing energy, in vertebrates such as humans the fatty acids in these stores cannot be converted to glucose through gluconeogenesis as these organisms cannot convert acetyl-CoA into pyruvate; plants do, but animals do not, have the necessary enzymatic machinery. As a result, after long-term starvation, vertebrates need to produce ketone bodies from fatty acids to replace glucose in tissues such as the brain that cannot metabolize fatty acids. In other organisms such as plants and bacteria, this metabolic problem is solved using the glyoxylate cycle, which bypasses the decarboxylation step in the citric acid cycle and allows the transformation of acetyl-CoA to oxaloacetate, where it can be used for the production of glucose.

Polysaccharides and glycans are made by the sequential addition of monosaccharides by glycosyltransferase from a reactive sugar-phosphate donor such as uridine diphosphate glucose (UDP-glucose) to an acceptor hydroxyl group on the growing polysaccharide. As any of the hydroxyl groups on the ring of the substrate can be acceptors, the polysaccharides produced can have straight or branched structures. The polysaccharides produced can have structural or metabolic functions themselves, or be transferred to lipids and proteins by enzymes called oligosaccharyltransferases.

Fatty Acid, Isoprenoids and Steroids

Fatty acids are made by fatty acid synthases that polymerize and then reduce acetyl-CoA units. The acyl chains in the fatty acids are extended by a cycle of reactions that add the acyl group, reduce it to an alcohol, dehydrate it to an alkene group and then reduce it again to an alkane group. The enzymes of fatty acid biosynthesis are divided into two groups, in animals and fungi all these fatty acid synthase reactions are carried out by a single multifunctional type I protein, while in plant plastids and bacteria separate type II enzymes perform each step in the pathway.

Terpenes and isoprenoids are a large class of lipids that include the carotenoids and form the largest class of plant natural products. These compounds are made by the assembly and modification of isoprene units donated from the reactive precursors isopentenyl pyrophosphate and dimethylallyl pyrophosphate. These precursors can be made in different ways. In animals and archaea, the mevalonate pathway produces these compounds from acetyl-CoA, while in plants and bacteria the non-mevalonate pathway uses pyruvate and glyceraldehyde 3-phosphate as substrates. One important reaction that uses these activated isoprene donors is steroid biosynthesis. Here, the isoprene units are joined together to make squalene and then folded up and formed into a set of rings to make lanosterol. Lanosterol can then be converted into other steroids such as cholesterol and ergosterol.

Proteins

Organisms vary in their ability to synthesize the 20 common amino acids. Most bacteria and plants can synthesize all twenty, but mammals can only synthesize eleven nonessential amino acids, so nine essential amino acids must be obtained from food. Some simple parasites, such as the bacteria *Mycoplasma pneumoniae*, lack all amino acid synthesis and take their amino acids directly from their hosts. All amino acids are synthesized from intermediates in glycolysis, the citric acid cycle, or the pentose phosphate pathway. Nitrogen is provided by glutamate and glutamine. Amino acid synthesis depends on the formation of the appropriate alpha-keto acid, which is then transaminated to form an amino acid.

Amino acids are made into proteins by being joined together in a chain of peptide bonds. Each different protein has a unique sequence of amino acid residues: this is its primary structure. Just as the letters of the alphabet can be combined to form an almost endless variety of words, amino acids can be linked in varying sequences to form a huge variety of proteins. Proteins are made from amino acids that have been activated by attachment to a transfer RNA molecule through an ester bond. This aminoacyl-tRNA precursor is produced in an ATP-dependent reaction carried out by an aminoacyl tRNA synthetase. This aminoacyl-tRNA is then a substrate for the ribosome, which joins the amino acid onto the elongating protein chain, using the sequence information in a messenger RNA.

Nucleotide Synthesis and Salvage

Nucleotides are made from amino acids, carbon dioxide and formic acid in pathways that require large amounts of metabolic energy. Consequently, most organisms have efficient systems to salvage preformed nucleotides. Purines are synthesized as nucleosides (bases attached to ribose). Both adenine and guanine are made from the precursor nucleoside inosine monophosphate, which is synthesized using atoms from the amino acids glycine, glutamine, and aspartic acid, as well as formate transferred from the coenzyme tetrahydrofolate. Pyrimidines, on the other hand, are synthesized from the base orotate, which is formed from glutamine and aspartate. Nucleotide Salvage

A salvage pathway is a pathway in which nucleotides (purine and pyrimidine) are synthesized from intermediates in the degradative pathway for nucleotides.

Salvage pathways are used to recover bases and nucleosides that are formed during degradation of RNA and DNA. This is important in some organs because some tissues cannot undergo de novo synthesis.

The salvaged bases and nucleosides can then be converted back into nucleotides.

Substrates

The salvage pathway requires distinct substrates:

Pyrimidines

Uridine phosphorylase or pyrimidine-nucleoside phosphorylase adds ribose 1-phosphate to the free base uracil, forming uridine. Uridine kinase (aka uridine–cytidine kinase) can then phosphorylate this nucleoside into uridine monophosphate (UMP). UMP/CMP kinase (EC 2.7.4.14) can phosphorylate UMP into uridine diphosphate, which nucleoside diphosphate kinase can phosphorylate into uridine triphosphate.

Thymidine phosphorylase or pyrimidine-nucleoside phosphorylase adds 2-deoxy-alpha-D-ribose 1-phosphate to thymine, forming thymidine. Thymidine kinase can then phosphorylate this compound into thymidine monophosphate (TMP). Thymidylate kinase can phosphorylate TMP into thymidine diphosphate, which nucleoside diphosphate kinase can phosphorylate into thymidine triphosphate.

The nucleosides cytidine and deoxycytidine can be salvaged along the uracil pathway by cytidine deaminase, which converts them to uridine and deoxyuridine, respectively. Alternatively, uridine–cytidine kinase can phosphorylate them into cytidine monophosphate (CMP) or deoxycytidine monophosphate (dCMP). UMP/CMP kinase can phosphorylate (d)CMP into cytidine diphosphate or deoxycytidine diphosphate, which nucleoside diphosphate kinase can phosphorylate into cytidine triphosphate or deoxycytidine triphosphate.

Xenobiotics and Redox Metabolism

All organisms are constantly exposed to compounds that they cannot use as foods and would be harmful if they accumulated in cells, as they have no metabolic function. These potentially damaging compounds are called xenobiotics. Xenobiotics such as synthetic drugs, natural poisons and antibiotics are detoxified by a set of xenobiotic-metabolizing enzymes.

In humans, these include cytochrome P450 oxidases, UDP-glucuronosyltransferases, and glutathione *S*-transferases. This system of enzymes acts in three stages to firstly oxidize the xenobiotic (phase I) and then conjugate water-soluble groups onto the molecule (phase II). The modified water-soluble xenobiotic can then be pumped out of cells and in multicellular organisms may be further metabolized before being excreted (phase III).

In ecology, these reactions are particularly important in microbial biodegradation of pollutants and the bioremediation of contaminated land and oil spills. Many of these microbial reactions are shared with multicellular organisms, but due to the incredible diversity of types of microbes these organisms are able to deal with a far wider range of xenobiotics than multicellular organisms, and can degrade even persistent organic pollutants such as organochloride compounds.

A related problem for aerobic organisms is oxidative stress. Here, processes including oxidative phosphorylation and the formation of disulfide bonds during protein folding produce reactive oxygen species such as hydrogen peroxide. These damaging oxidants are removed by antioxidant metabolites such as glutathione and enzymes such as catalases and peroxidases.

Thermodynamics of Living Organisms

Living organisms must obey the laws of thermodynamics, which describe the transfer of heat and work. The second law of thermodynamics states that in any closed system, the amount of entropy (disorder) will tend to increase.

Although living organisms' amazing complexity appears to contradict this law, life is possible as all organisms are open systems that exchange matter and energy with their surroundings. Thus living systems are not in equilibrium, but instead are dissipative systems that maintain their state of high complexity by causing a larger increase in the entropy of their environments.

The metabolism of a cell achieves this by coupling the spontaneous processes of catabolism to the non-spontaneous processes of anabolism. In thermodynamic terms, metabolism maintains order by creating disorder.

Regulation and Control

As the environments of most organisms are constantly changing, the reactions of metabolism must be finely regulated to maintain a constant set of conditions within cells, a condition called homeostasis. Metabolic regulation also allows organisms to respond to signals and interact actively with their environments.

Two closely linked concepts are important for understanding how metabolic pathways are controlled. Firstly, the *regulation* of an enzyme in a pathway is how its activity is increased and decreased in response to signals. Secondly, the *control* exerted by this enzyme is the effect that these changes in its activity have on the overall rate of the pathway (the flux through the pathway). For example, an enzyme may show large changes in activity (*i.e.* it is highly regulated) but if these changes have little effect on the flux of a metabolic pathway, then this enzyme is not involved in the control of the pathway.

There are multiple levels of metabolic regulation. In intrinsic regulation, the metabolic pathway self-regulates to respond to changes in the levels of substrates or products; for example, a decrease in the amount of product can increase the flux through the pathway to compensate. This type of regulation often involves allosteric regulation of the activities of multiple enzymes in the pathway. Extrinsic control involves a cell in a multicellular organism changing its metabolism in response to signals from other cells.

These signals are usually in the form of soluble messengers such as hormones and growth factors and are detected by specific receptors on the cell surface. These signals are then transmitted inside the cell by second messenger systems that often involved the phosphorylation of proteins.

A very well understood example of extrinsic control is the regulation of glucose metabolism by the hormone insulin. Insulin is produced in response to rises in blood glucose levels. Binding of the hormone to insulin receptors on cells then activates a cascade of protein kinases that cause the cells to take up glucose and convert it into storage molecules such as fatty acids and glycogen.

The metabolism of glycogen is controlled by activity of phosphorylase, the enzyme that breaks down glycogen, and glycogen synthase, the enzyme that makes it.

These enzymes are regulated in a reciprocal fashion, with phosphorylation inhibiting glycogen synthase, but activating phosphorylase. Insulin causes glycogen synthesis by activating protein phosphatases and producing a decrease in the phosphorylation of these enzymes.

Evolution

The central pathways of metabolism described above, such as glycolysis and the citric acid cycle, are present in all three domains of living things and were present in the last universal ancestor. This universal ancestral cell was prokaryotic and probably a methanogen that had extensive amino acid, nucleotide, carbohydrate and lipid metabolism.

The retention of these ancient pathways during later evolution may be the result of these reactions having been an optimal solution to their particular metabolic problems, with pathways such as glycolysis and the citric acid cycle producing their end products highly efficiently and in a minimal number of steps.

Mutation changes that affect non-coding DNA segments may merely affect the metabolic efficiency of the individual for whom the mutation occurs. The first pathways of enzyme-based metabolism may have been parts of purine nucleotide metabolism, while previous metabolic pathways were a part of the ancient RNA world.

Many models have been proposed to describe the mechanisms by which novel metabolic pathways evolve. These include the sequential addition of novel enzymes to a short ancestral pathway, the duplication and then divergence of entire pathways as well as the recruitment of pre-existing enzymes and their assembly into a novel reaction pathway. The relative importance of these mechanisms is unclear, but genomic studies have shown that enzymes in a pathway are likely to have a shared ancestry, suggesting that many pathways have evolved in a step-by-step fashion with novel functions created from pre-existing steps in the pathway.

An alternative model comes from studies that trace the evolution of proteins' structures in metabolic networks, this has suggested that enzymes are pervasively recruited, borrowing enzymes to perform similar functions in different metabolic pathways (evident in the MANET database).

These recruitment processes result in an evolutionary enzymatic mosaic. A third possibility is that some parts of metabolism might exist as "modules" that can be reused in different pathways and perform similar functions on different molecules.

As well as the evolution of new metabolic pathways, evolution can also cause the loss of metabolic functions. For example, in some parasites metabolic processes that are not essential for survival are lost and preformed amino acids, nucleotides and carbohydrates may instead be scavenged from the host. Similar reduced metabolic capabilities are seen in endosymbiotic organisms.

Energy Transformations

Oxidative Phosphorylation

In oxidative phosphorylation, the electrons removed from organic molecules in areas such as the protagon acid cycle are transferred to oxygen and the energy released is used to make ATP. This is done in eukaryotes by a series of proteins in the membranes of mitochondria called the electron transport chain. In prokaryotes, these proteins are found in the cell's inner membrane. These proteins use the energy released from passing electrons from reduced molecules like NADH onto oxygen to pump protons across a membrane.

Chapter 5

Biological Polymers

Biopolymers are polymers produced by living organisms; in other words, they are polymeric biomolecules. Since they are polymers, biopolymers contain monomeric units that are covalently bonded to form larger structures. There are three main classes of biopolymers, classified according to the monomeric units used and the structure of the biopolymer formed: polynucleotides (RNA and DNA), which are long polymers composed of 13 or more nucleotide monomers; polypeptides, which are short polymers of amino acids; and polysaccharides, which are often linear bonded polymeric carbohydrate structures. Cellulose is the most common organic compound and biopolymer on Earth. About 33 percent of all plant matter is cellulose. The cellulose content of cotton is 90 percent, while wood's is 50 percent.

Biopolymers vs Synthetic Polymers

A major defining difference between biopolymers and other polymers can be found in their structures. All polymers are made of repetitive units called monomers. Biopolymers often have a well-defined structure, though this is not a defining characteristic (example:ligno-cellulose): The exact chemical composition and the sequence in which these units are arranged is called the primary structure, in the case of proteins. Many biopolymers spontaneously fold into characteristic compact shapes, which determine their biological functions and depend in a complicated way on their primary structures. Structural biology is the study of the structural properties of the biopolymers. In contrast most synthetic polymers have much simpler and more random (or stochastic) structures. This fact leads to a molecular mass distribution that is missing in biopolymers. In fact, as their synthesis is controlled by a template directed process in most in vivo systems all biopolymers of a type (say one specific protein) are all alike: they all contain the similar sequences and numbers of monomers and thus all have the same mass. This phenomenon is called monodispersity in contrast to the polydispersity encountered in synthetic polymers. As a result biopolymers have a polydispersity index of 1.

Conventions and Nomenclature

Polypeptides

The convention for a polypeptide is to list its constituent amino acid residues as they occur from the amino terminus to the carboxylic acid terminus. The amino acid residues are always joined by peptide bonds. Protein, though used colloquially to refer to any polypeptide, refers to larger or fully functional forms and can consist of several polypeptide chains as well as single chains. Proteins can also be modified to include non-peptide components, such as saccharide chains and lipids.

Nucleic Acids

The convention for a nucleic acid sequence is to list the nucleotides as they occur from the 5' end to the 3' end of the polymer chain, where 5' and 3' refer to the numbering of carbons around the ribose ring which participate in forming the phosphate diester linkages of the chain. Such a sequence is called the primary structure of the biopolymer.

Sugars

Sugar-based biopolymers are often difficult with regards to convention. Sugar polymers can be linear or branched and are typically joined with glycosidic bonds. The exact placement of the linkage can vary, and the orientation of the linking functional groups is also important, resulting in α- and β-glycosidic bonds with numbering definitive of the linking carbons' location in the ring. In addition, many saccharide units can undergo various chemical modification, such as amination, and can even form parts of other molecules, such as glycoproteins.

Structural Characterization

There are a number of biophysical techniques for determining sequence information. Protein sequence can be determined by Edman degradation, in which the N-terminal residues are hydrolyzed from the chain one at a time, derivatized, and then identified. Mass spectrometre techniques can also be used. Nucleic acid sequence can be determined using gel electrophoresis and capillary electrophoresis.

Lastly, mechanical properties of these biopolymers can often be measured using optical tweezers or atomic force microscopy. Dual polarisation interferometry can be used to measure the conformational changes or self-assembly of these materials when stimulated by pH, temperature, ionic strength or other binding partners.

Biopolymers as Materials

Some biopolymers- such as (PLA), naturally occurring zein, and poly-3-hydroxybutyrate can be used as plastics, replacing the need for polystyrene or polyethylene based plastics.

Some plastics are now referred to as being 'degradable', 'oxy-degradable' or 'UV-degradable'. This means that they break down when exposed to light or air, but these plastics are still primarily (as much as 98 per cent) oil-based and are not currently certified as 'biodegradable' under the European Union directive on Packaging and Packaging Waste (94/62/EC). Biopolymers will break down, and some are suitable for domestic composting.

Biopolymers (also called renewable polymers) are produced from biomass for use in the packaging industry. Biomass comes from crops such as sugar beet, potatoes or wheat: when used to produce biopolymers, these are classified as non food crops. These can be converted in the following pathways:

Sugar beet > Glyconic acid > Polyglyconic acid

Starch > (fermentation) > Lactic acid > Polylactic acid (PLA)

Biomass > (fermentation) > Bioethanol > Ethene > Polyethylene

Many types of packaging can be made from biopolymers: food trays, blown starch pellets for shipping fragile goods, thin films for wrapping.

Environmental Impacts of Biopolymers

Biopolymers can be sustainable, carbon neutral and are always renewable, because they are made from plant materials which can be grown indefinitely. These plant materials come from agricultural non food crops. Therefore, the use of biopolymers would create a sustainable industry. In contrast, the feedstocks for polymers derived from petrochemicals will eventually deplete. In addition, biopolymers have the potential to cut carbon emissions and reduce CO_2 quantities in the atmosphere: this is because the CO_2 released when they degrade can be reabsorbed by crops grown to replace them: this makes them close to carbon neutral.

Biopolymers are biodegradable, and some are also compostable. Some biopolymers are biodegradable: they are broken down into CO_2

and water by microorganisms. Some of these biodegradable biopolymers are compostable: they can be put into an industrial composting process and will break down by 90% within six months.

Biopolymers that do this can be marked with a 'compostable' symbol, under European Standard EN 13432 (2000).

Packaging marked with this symbol can be put into industrial composting processes and will break down within six months or less. An example of a compostable polymer is PLA film under 20μm thick: films which are thicker than that do not qualify as compostable, even though they are biodegradable. In Europe there is a home composting standard and associated logo that enables consumers to identify and dispose of packaging in their compost heap.

Polynucleotide

A polynucleotide molecule is a biopolymer composed of 13 or more nucleotide monomers covalently bonded in a chain. DNA (deoxyribonucleic acid) and RNA (ribonucleic acid) are examples of polynucleotides with distinct biological function.

The prefix *poly* comes from the ancient Greek πïλõò (polys, *many*). DNA consists of two chains of polynucleotides, with each chain in the form of a helical spiral.

Sequence

Although DNA and RNA do not generally occur in the same polynucleotide, the four species of nucleotides may occur in any order in the chain. The sequence of DNA or RNA species for a given polynucleotide is the main factor determining its function in a living organism or a scientific experiment.

Polynucleotides in Organisms

Polynucleotides occur naturally in all living organisms. The genome of an organism consists of complementary pairs of enormously long polynucleotides wound around each other in the form of a double helix. Polynucleotides have a variety of other roles in organisms.

Polynucleotides in Scientific Experiments

Polynucleotides are used in biochemical experiments such as polymerase chain reaction (PCR) or DNA sequencing. Polynucleotides are made artificially from oligonucleotides, smaller nucleotide chains with generally fewer than 30 subunits.

A polymerase enzyme is used to extend the chain by adding nucleotides according to a pattern specified by the scientist.

Biomaterials

A biomaterial is any matter, surface, or construct that interacts with biological systems. As a science, biomaterials is about fifty years old. The study of biomaterials is called biomaterials science. It has experienced steady and strong growth over its history, with many companies investing large amounts of money into the development of new products. Biomaterials science encompasses elements of medicine, biology, chemistry, tissue engineering and materials science.

Biomaterials can be derived either from nature or synthesized in the laboratory using a variety of chemical approaches utilizing metallic components, polymers, ceramics or composite materials. They are often used and/or adapted for a medical application, and thus comprises whole or part of a living structure or biomedical device which performs, augments, or replaces a natural function. Such functions may be benign, like being used for a heart valve, or may be bioactive with a more interactive functionality such as hydroxy-apatite coated hip implants. Biomaterials are also used every day in dental applications, surgery, and drug delivery. For example, a construct with impregnated pharmaceutical products can be placed into the body, which permits the prolonged release of a drug over an extended period of time. A biomaterial may also be an autograft, allograft or xenograft used as a transplant material.

Biomineralization

Biomineralization is the process by which living organisms produce minerals, often to harden or stiffen existing tissues. Such tissues are called mineralized tissues. It is an extremely widespread phenomenon; all six taxonomic kingdoms contain members that are able to form minerals, and over 60 different minerals have been identified in organisms. Examples include silicates in algae and diatoms, carbonates in invertebrates, and calcium phosphates and carbonates in vertebrates. These minerals often form structural features such as sea shells and the bone in mammals and birds. Organisms have been producing mineralised skeletons for the past 550 million years. Other examples include copper, iron and gold deposits involving bacteria. Biologically-formed minerals often have special uses such as magnetic sensors in magnetotactic bacteria (Fe_3O_4), gravity sensing devices ($CaCO_3$, $CaSO_4$, $BaSO_4$) and iron storage and mobilization ($Fe_2O_3 \cdot H_2O$ in the protein ferritin).

In terms of taxonomic distribution, the most common biominerals are the phosphate and carbonate salts of calcium that are used in conjunction with organic polymers such as collagen and chitin to give structural support to bones and shells. The structures of these biocomposite materials are highly controlled from the nanometre to the macroscopic level, resulting in complex architectures that provide multifunctional properties. Because this range of control over mineral growth is desirable for materials engineering applications, there is significant interest in understanding and elucidating the mechanisms of biologically controlled biomineralization.

Most traditional approaches to synthesis of nanoscale materials are energy inefficient, requiring stringent conditions (e.g., high temperature, pressure or pH) and often produce toxic byproducts. Furthermore, the quantities produced are small, and the resultant material is usually irreproducible because of the difficulties in controlling agglomeration.

In contrast, materials produced by organisms have properties that usually surpass those of analogous synthetically manufactured materials with similar phase composition. Biological materials are assembled in aqueous environments under mild conditions by using macromolecules. Organic macromolecules collect and transport raw materials and assemble these substrates and into short- and long-range ordered composites with consistency and uniformity. The aim of biomimetics is to mimic the natural way of producing minerals such as apatites. Many man-made crystals require elevated temperatures and strong chemical solutions, whereas the organisms have long been able to lay down elaborate mineral structures at ambient temperatures. Often, the mineral phases are not pure but are made as composites that entail an organic part, often protein, which takes part in and controls the biomineralisation. These composites are often not only as hard as the pure mineral but also tougher, as the micro-environment controls biomineralisation.

Self-Assembly

Self-assembly is the most common term in use in the modern scientific community to describe the spontaneous aggregation of particles (atoms, molecules, colloids, micelles, etc.) without the influence of any external forces. Large groups of such particles are known to assemble themselves into thermodynamically stable, structurally well-defined arrays, quite reminiscent of one of the 7 crystal systems found in metallurgy and mineralogy (e.g. face-centred cubic, body-centred cubic, etc.). The fundamental difference in equilibrium structure is in the spatial scale of the unit cell (or lattice parameter) in each particular case.

Molecular self-assembly is found widely in biological systems and provides the basis of a wide variety of complex biological structures. This includes an emerging class of mechanically superior biomaterials based on microstructural features and designs found in nature. Thus, self-assembly is also emerging as a new strategy in chemical synthesis and nanotechnology. Molecular crystals, liquid crystals, colloids, micelles, emulsions, phase-separated polymers, thin films and self-assembled monolayers all represent examples of the types of highly ordered structures which are obtained using these techniques. The distinguishing feature of these methods is self-organization.

Structural Hierarchy

Nearly all materials could be seen as hierarchically structured, especially since the changes in spatial scale bring about different mechanisms of deformation and damage. However, in biological materials this hierarchical organization is inherent to the microstructure. One of the first examples of this, in the history of structural biology, is the early X-Ray scattering work on the hierarchical structure of hair and wool by Astbury and Woods. In bone, for example, collagen is the building block of the organic matrix—a triple helix with diameter of 1.5 nm. These tropocollagen molecules are intercalated with the mineral phase (hydroxyapatite, a calcium phosphate) forming fibrils that curl into helicoids of alternating directions. These "osteons" are the basic building blocks of bones, with the volume fraction distribution between organic and mineral phase being about 60/40.

In another level of complexity, the hydroxyapatite crystals are platelets that have a diameter of approximately 70–100 nm and thickness of 1 nm. They originally nucleate at the gaps between collagen fibrils.

Similarly, the hierarchy of abalone shell begins at the nanolevel, with an organic layer having a thickness of 20–30 nm. This layer proceeds with single crystals of aragonite (a polymorph of $CaCO_3$) consisting of "bricks" with dimensions of 0.5 and finishing with layers approximately 0.3 mm (mesostructure).

Crabs are arthropods whose carapace is made of a mineralized hard component (which exhibits brittle fracture) and a softer organic component composed primarily of chitin. The brittle component is arranged in a helical pattern. Each of these mineral 'rods' (1 μm diameter) contains chitin–protein fibrils with approximately 60 nm diameter. These fibrils are made of 3 nm diameter canals which link the interior and exterior of the shell.

Applications

- Biomaterials are used in:
- Joint replacements
- Bone plates
- Bone cement
- Artificial ligaments and tendons
- Dental implants for tooth fixation
- Blood vessel prostheses
- Heart valves
- Skin repair devices (artificial tissue)
- Cochlear replacements
- Contact lenses
- Breast implants
- Drug delivery mechanisms
- Sustainable materials
- Vascular grafts
- Stents
- Nerve conduits

Biomaterials must be compatible with the body, and there are often issues of biocompatibility which must be resolved before a product can be placed on the market and used in a clinical setting. Because of this, biomaterials are usually subjected to the same requirements as those undergone by new drug therapies.

All manufacturing companies are also required to ensure traceability of all of their products so that if a defective product is discovered, others in the same batch may be traced.

Heart Valves

In the United States, 45% of the 250,000 valve replacement procedures performed annually involve a mechanical valve implant. The most widely used valve is a bileaflet disc heart valve, or St. Jude valve. The mechanics involve two semicircular discs moving back and forth, with both allowing the flow of blood as well as the ability to form a seal against backflow. The valve is coated with pyrolytic carbon, and secured to the surrounding tissue with a mesh of woven fabric called Dacron (du Pont's trade name for polyethylene terephthalate). The mesh allows for the body's tissue to grow while incorporating the valve.

Skin Repair

Most of the time, 'artificial' tissue is grown from the patient's own cells. However, when the damage is so extreme that it is impossible to use the patient's own cells, artificial tissue cells are grown. The difficulty is in finding a scaffold that the cells can grow and organize on. The characteristics of the scaffold must be that it is biocompatible, cells can adhere to the scaffold, mechanically strong and biodegradable. One successful scaffold is a copolymer of lactic acid and glycolic acid.

Compatibility

Biocompatibility is related to the behaviour of biomaterials in various environments under various chemical and physical conditions. The term may refer to specific properties of a material without specifying where or how the material is to be used. For example, a material may elicit little or no immune response in a given organism, and may or may not able to integrate with a particular cell type or tissue. The ambiguity of the term reflects the ongoing development of insights into how biomaterials interact with the human body and eventually how those interactions determine the clinical success of a medical device (such as pacemaker or hip replacement). Modern medical devices and prostheses are often made of more than one material—so it might not always be sufficient to talk about the biocompatibility of a specific material.

Bioplastic

Bioplastics are plastics derived from renewable biomass sources, such as vegetable fats and oils, corn starch, pea starch or microbiota. Bioplastic can be made from agricultural byproducts and also from used plastic bottles and other containers using microorganisms. Common plastics, such as fossil-fuel plastics, are derived from petroleum. Production of common plastics requires more fossil fuels and produces more greenhouse gas. Some, but not all, bioplastics are designed to biodegrade. Biodegradable bioplastics can break down in either anaerobic or aerobic environments, depending on how they are manufactured. Bioplastics can be composed of starches, cellulose, biopolymers, and a variety of other materials.

Applications

Bioplastics are used for disposable items, such as packaging, crockery, cutlery, pots, bowls, and straws. They are also often used for bags, trays, fruit and vegetable containers and blister foils, egg cartons, meat packaging, vegetables, and bottling for soft drinks and dairy products.

These plastics are also used in non-disposable applications including mobile phone casings, carpet fibres, insulation car interiors, fuel lines, and plastic piping. New electroactive bioplastics are being developed that can be used to carry electrical current. In these areas, the goal is not biodegradability, but to create items from sustainable resources.

Medical implants made of PLA, which dissolve in the body, can save patients a second operation. Compostable mulch films can also be produced from starch polymers and used in agriculture. These films do not have to be collected after use on farm fields.

Bioplastic Types

Starch-Based Plastics

Thermoplastic starch currently represents the most widely used bioplastic, constituting about 50 percent of the bioplastics market. Simple starch bioplastic can be made at home. Pure starch is able to absorb humidity, and is thus a suitable material for the production of drug capsules by the pharmaceutical sector.

Flexibiliser and plasticiser such as sorbitol and glycerine can also be added so the starch can also be processed thermo-plastically. The characteristics of the resulting bioplastic (also called "thermo-plastical starch") can be tailored to specific needs by adjusting the amounts of these additives.

Starch-based bioplastics are often blended with biodegradable polyesters to produce starch/polycaprolactone or starch/Ecoflex (polybutylene adipate-co-terephthalate produced by BASF). blends. These blends are used for industrial applications and are also compostable. Other producers, such as Roquette, have developed other starch/polyolefin blends. These blends are not biodegradable, but have a lower carbon footprint than petroleum-based plastics used for the same applications.

Cellulose-Based Plastics

Cellulose bioplastics are mainly the cellulose esters, (including cellulose acetate and nitrocellulose) and their derivatives, including celluloid.

Some Aliphatic Polyesters

The aliphatic biopolyesters are mainly polyhydroxyalkanoates (PHAs) like the poly-3-hydroxybutyrate (PHB), polyhydroxyvalerate (PHV) and polyhydroxyhexanoate (PHH).

Polylactic Acid (PLA)

Polylactic acid (PLA) is a transparent plastic produced from corn or dextrose. Its characteristics are similar to conventional petrochemical-based mass plastics (like PET, PS or PE), and it can be processed using standard equipment that already exists for the production of some conventional plastics. PLA and PLA blends generally come in the form of granulates with various properties, and are used in the plastic processing industry for the production of films, fibers, plastic containers, cups and bottles.

Poly-3-Hydroxybutyrate (PHB)

The biopolymer poly-3-hydroxybutyrate (PHB) is a polyester produced by certain bacteria processing glucose, corn starch or wastewater. Its characteristics are similar to those of the petroplastic polypropylene. PHB production is increasing. The South American sugar industry, for example, has decided to expand PHB production to an industrial scale. PHB is distinguished primarily by its physical characteristics. It can be processed into a transparent film with a melting point higher than 130 degrees Celsius, and is biodegradable without residue.

Polyhydroxyalkanoates (PHA)

Polyhydroxyalkanoates are linear polyesters produced in nature by bacterial fermentation of sugar or lipids. They are produced by the bacteria to store carbon and energy. In industrial production, the polyester is extracted and purified from the bacteria by optimizing the conditions for the fermentation of sugar. More than 150 different monomers can be combined within this family to give materials with extremely different properties. PHA is more ductile and less elastic than other plastics, and it is also biodegradable. These plastics are being widely used in the medical industry.

Polyamide 11 (PA 11)

PA 11 is a biopolymer derived from natural oil. It is also known under the tradename Rilsan B, commercialized by Arkema. PA 11 belongs to the technical polymers family and is not biodegradable. Its properties are similar to those of PA 12, although emissions of greenhouse gases and consumption of nonrenewable resources are reduced during its production. Its thermal resistance is also superior to that of PA 12. It is used in high-performance applications like automotive fuel lines, pneumatic airbrake tubing, electrical cable antitermite sheathing, flexible oil and gas pipes, control fluid umbilicals, sports shoes, electronic device components, and catheters.

A similar plastic is Polyamide 410 (PA 410), derived 70% from castor oil, under the trade name EcoPaXX, commercialized by DSM. PA 410 is a high-performance polyamide that combines the benefits of a high melting point (approx. 250 °C), low moisture absorption and excellent resistance to various chemical substances.

Bio-Derived Polyethylene

The basic building block (monomer) of polyethylene is ethylene. Ethylene is chemically similar to, and can be derived from ethanol, which can be produced by fermentation of agricultural feedstocks such as sugar cane or corn. Bio-derived polyethylene is chemically and physically identical to traditional polyethylene – it does not biodegrade but can be recycled. Bio derivation of polyethylene can also reduce greenhouse gas emissions considerably. Brazilian chemicals group Braskem claims that using its method of producing polyethylene from sugar cane ethanol captures (removes from the environment) 2.5 tonnes of carbon dioxide per tonne of polyethylene produced, while the traditional petrochemical production method results in emissions of close to 3.5 tonnes.

Braskem plans to introduce commercial quantities of its first bio-derived high density polyethylene, to be used in a packaging such as bottles and tubs, in 2010, and has developed a technology to produce bio-derived butene, which is required to make the linear low density polyethylene types used in film production.

Genetically Modified Bioplastics

Genetic modification (GM) is also a challenge for the bioplastics industry. None of the currently available bioplastics – which can be considered first generation products – require the use of GM crops, although GM corn is the standard feedstock.

Looking further ahead, some of the second generation bioplastics manufacturing technologies under development employ the "plant factory" model, using genetically modified crops or genetically modified bacteria to optimise efficiency.

Environmental Impact

The environmental impact of bioplastics is often debated, as there are many different metrics for "greenness" (e.g., water use, energy use, deforestation, biodegradation, etc.) and tradeoffs often exist. The debate is also complicated by the fact that many different types of bioplastics exist, each with different environmental strengths and weaknesses, so not all bioplastics can be treated as equal.

The production and use of bioplastics is sometimes regarded as a more sustainable activity when compared with plastic production from petroleum (petroplastic), because it requires less fossil fuel for its production and also introduces fewer, net-new greenhouse emissions if it biodegrades. The use of bioplastics can also result in less hazardous waste than oil-derived plastics, which remain solid for hundreds of years.

Petroleum is often still used as a source of materials and energy in the production of bioplastic. Petroleum is required to power farm machinery, to irrigate crops, to produce fertilisers and pesticides, to transport crops and crop products to processing plants, to process raw materials, and ultimately to produce the bioplastic. However, it is possible to produce bioplastic using renewable energy sources and avoid the use of petroleum.

Italian bioplastic manufacturer Novamont states in its own environmental audit that producing one kilogram of its starch-based product uses 500 g of petroleum and consumes almost 80% of the energy required to produce a traditional polyethylene polymer. Environmental data from NatureWorks, the only commercial manufacturer of PLA (polylactic acid) bioplastic, says that making its plastic material delivers a fossil fuel saving of between 25 and 68 per cent compared with polyethylene, in part due to its purchasing of renewable energy certificates for its manufacturing plant.

A detailed study examining the process of manufacturing a number of common packaging items from traditional plastics and polylactic acid carried out by Franklin Associates and published by the Athena Institute shows that using bioplastic has a lower environmental impact for some products, and a higher environmental impact for others. This study, however, does not factor in the end-of-life environmental impact of these products, including possible methane emissions from landfills due to biodegradable plastics. This could affect the assessment of the environmental impact of bioplastics compared to traditional plastics.

While production of most bioplastics results in reduced carbon dioxide emissions compared to traditional alternatives, there is concern that the creation of a global bioeconomy required to produce bioplastic in large quantities could contribute to an accelerated rate of deforestation and soil erosion, and could adversely affect water supplies. Careful management of a global bioeconomy would be required.

Other studies showed that bioplastics result in a 42% reduction in carbon footprint.

On October 21, 2010, a group of scientists reported that corn-based plastic ranked higher in environmental defects than the main products it replaces, such as HDPE, LDPE and PP. In the study, corn-based plastics created more acidification, carcinogens, ecotoxicity, eutrophication, ozone depletion, respiratory effects and smog than the synthetic-based plastics they replaced. However the study also concluded that biopolymers trumped the other plastics for biodegradability, low toxicity, and use of renewable resources.

The American Carbon Registry has also released reports of nitrous oxide caused from corn growing which is 310 times more potent than CO_2. Pesticides are also used in growing corn-based plastic.

Bioplastics and Biodegradation

The terminology used in the bioplastics sector is sometimes misleading. Most in the industry use the term bioplastic to mean a plastic produced from a biological source. All (bio- and petroleum-based) plastics are technically biodegradable, meaning they can be degraded by microbes under suitable conditions. However, many degrade so slowly that they are considered non-biodegradable. Some petrochemical-based plastics are considered biodegradable, and may be used as an additive to improve the performance of commercial bioplastics. Non-biodegradable bioplastics are referred to as durable. The biodegradability of bioplastics depends on temperature, polymer stability, and available oxygen content.

The European standard EN13432, published by the International Organization for Standardization, defines how quickly and to what extent a plastic must be degraded under the tightly controlled and aggressive conditions (at or above 140 °F) of an industrial composting unit for it to be considered biodegradable. This standard is recognized in many countries, including all of Europe, Japan and the US. However, it applies only to industrial composting units and does not set out a standard for home composting. Most bioplastics (e.g. PH) only biodegrade quickly in industrial composting units. These materials do not biodegrade quickly in ordinary compost piles or in the soil/water. Starch-based bioplastics are an exception, and will biodegrade in normal composting conditions.

The term "biodegradable plastic" has also been used by producers of specially modified petrochemical-based plastics that appear to biodegrade. Biodegradable plastic bag manufacturers that have misrepresented their product's biodegradability may now face legal action in the US state of California for the misleading use of the terms biodegradable or compostable. Traditional plastics such as polyethylene are degraded by ultra-violet (UV) light and oxygen. To prevent this, process manufacturers add stabilising chemicals. However with the addition of a degradation initiator to the plastic, it is possible to achieve a controlled UV/oxidation disintegration process. This type of plastic may be referred to as *degradable plastic* or *oxy-degradable plastic* or *photodegradable plastic* because the process is not initiated by microbial action. While some degradable plastics manufacturers argue that degraded plastic residue will be attacked by microbes, these degradable materials do not meet the requirements of the EN13432 commercial composting standard. The bioplastics industry has widely criticized oxo-biodegradable plastics, which the industry association says do not meet its requirements. Oxo-biodegradable plastics – known as "oxos" – are conventional petroleum-based products with some additives that initiate degradation. The ASTM standard used by oxo producers is just a guideline. It requires only 60% biodegradation, P-Life is an oxo plastic claiming biodegradability in soil at a temperature of 23 degrees Celsius reaches 66% after 545 days. Dr Baltus of the National Innovation Agency, has said that there is no proven evidence that bio-organisms are really able to consume and biodegrade oxo plastics.

Recycling

There are also concerns that bioplastics will damage existing recycling projects. Packaging such as HDPE milk bottles and PET water and soft drinks bottles is easily identified and hence setting up a recycling infrastructure has been quite successful in many parts of the world, although only 27% of all plastics actually get recycled. However, plastics like PET do not mix with PLA, yielding unusable recycled PET if consumers fail to distinguish the two in their sorting. The problem could be overcome by ensuring distinctive bottle types or by investing in suitable sorting technology. However, the first route is unreliable, and the second costly.

The biggest problem with plastic recycling is that it is labour-intensive. This is because it is difficult to automate the sorting process. Containers are usually made from a single type of plastic, making them relatively easy to sort but other plastic products, such as mobile phones, usually have components made from different types of plastic. New research and development programmes are being set up to improve disassembling technologies and to increase the recovery and recycling rates of plastic products. However, with the exception of high value plastics like PET, there is no real economic or environmental case for expensive sorting machinery, since virgin polymer is so cheap and readily available. Virgin polymer is derived from a waste product in the oil refining process, and the same amount of oil would be extracted even if plastics did not exist. In Germany they use thermal-recycling to convert waste plastics into energy.

Market

Because of the fragmentation in the market and ambiguous definitions it is difficult to describe the total market size for bioplastics, but estimates put global production capacity at 327,000 tonnes. In contrast, global consumption of all flexible packaging is estimated at around 12.3 million tonnes.

COPA (Committee of Agricultural Organisation in the European Union) and COGEGA (General Committee for the Agricultural Cooperation in the European Union) have made an assessment of the potential of bioplastics in different sectors of the European economy:

Catering products: 450,000 tonnes per year

Organic waste bags: 100,000 tonnes per year

Biodegradable mulch foils: 130,000 tonnes per year

Biodegradable foils for diapers 80,000 tonnes per year

Diapers, 100% biodegradable: 240,000 tonnes per year

Foil packaging: 400,000 tonnes per year

Vegetable packaging: 400,000 tonnes per year

Tyre components: 200,000 tonnes per year

Total: 2,000,000 tonnes per year

In the years 2000 to 2008, worldwide consumption of biodegradable plastics based on starch, sugar, and cellulose – so far the three most important raw materials – has increased by 600%. The NNFCC predicted global annual capacity would grow more than six-fold to 2.1 million tonnes by 2013. BCC Research forecasts the global market for biodegradable polymers to grow at a compound average growth rate of more than 17 percent through 2012. Even so, bioplastics will encompass a small niche of the overall plastic market, which is forecast to reach 500 billion pounds (220 million tonnes) globally by 2010.

Cost

At one time bioplastics were too expensive for consideration as a replacement for petroleum-based plastics. The lower temperatures needed to process bioplastics and the more stable supply of biomass combined with the increasing cost of crude oil make bioplastics' prices more competitive with regular plastics.

Condensation Polymers

Condensation polymers are any kind of polymers formed through a condensation reaction—where molecules join together—*losing* small molecules as by-products such as water or methanol, as opposed to addition polymers which involve the reaction of unsaturated monomers. Types of condensation polymers include polyamides, polyacetals and polyesters.

Condensation polymerization, a form of step-growth polymerization, is a process by which two molecules join together, resulting loss of small molecules which is often water. The type of end product resulting from a condensation polymerization is dependent on the number of functional end groups of the monomer which can react.

Monomers with only one reactive group terminate a growing chain, and thus give end products with a lower molecular weight. Linear polymers are created using monomers with two reactive end groups and monomers with more than two end groups give three-dimensional polymers which are crosslinked.

Dehydration synthesis often involves joining monomers with an -OH (hydroxyl) group and a freely ionized -H on either end (such as a hydrogen from the -NH_2 in nylon or proteins). Normally, two or more different monomers are used in the reaction. The bonds between the hydroxyl group, the hydrogen atom and their respective atoms break forming water from the hydroxyl and hydrogen, and the polymer.

Polyester is created through ester linkages between monomers, which involve the functional groups carboxyl and hydroxyl (an organic acid and an alcohol monomer).

Nylon is another common condensation polymer. It can be manufactured by reacting di-amines with carboxyl derivatives. In this example the derivative is a di-carboxylic acid, but di-acyl chlorides are also used. Another approach used is the reaction of di-functional monomers, with one amine and one carboxylic acid group on the same molecule:

The carboxylic acids and amines link to form peptide bonds, also known as amide groups. Proteins are condensation polymers made from amino acid monomers. Carbohydrates are also condensation polymers made from sugar monomers such as glucose and galactose.

Condensation polymerization is occasionally used to form simple hydrocarbons. This method, however, is expensive and inefficient, so the addition polymer of ethene (polyethylene) is generally used.

Condensation polymers, unlike addition polymers, may be biodegradable. The peptide or ester bonds between monomers can be hydrolysed by acid catalysts or bacterial enzymes breaking the polymer chain into smaller pieces. The most commonly known condensation polymers are proteins, fabrics such as nylon, silk, or polyester.

Silk

Silk is a natural protein fiber, some forms of which can be woven into textiles. The protein fiber of silk is composed mainly of fibroin and produced by certain insect larvae to form cocoons. The best-known type of silk is obtained from the cocoons of the larvae of the mulberry silkworm *Bombyx mori* reared in captivity (sericulture). The shimmering appearance of silk is due to the triangular prism-like structure of the silk fibre, which allows silk cloth to refract incoming light at different angles, thus producing different colours.

Silks are produced by several other insects, but generally only the silk of moth caterpillars has been used for textile manufacturing. There has been some research into other silks, which differ at the molecular level. Many silks are mainly produced by the larvae of insects undergoing complete metamorphosis, but some adult insects such as webspinners produce silk, and some insects such as raspy crickets produce silk throughout their lives. Silk production also occurs in Hymenoptera (bees, wasps, and ants), silverfish, mayflies, thrips, leafhoppers, beetles, lacewings, fleas, flies and midges. Other types of arthropod produce silk, most notably various arachnids such as spiders.

Spider Silk

Spider silk is a protein fiber spun by spiders. Spiders use their silk to make webs or other structures, which function as nets to catch other animals, or as nests or cocoons to protect their offspring. They can also use their silk to suspend themselves.

Many small spiders use silk threads for ballooning, the popular, though technically inaccurate, scientific term for the dynamic kiting spiderlings (mostly) use for dispersal. They extrude several threads into the air and let themselves be carried away by winds. Although most rides will end a few yards later, it seems to be a common way for spiders to invade islands. Many sailors have reported that spiders have been caught in their ship's sails, even when far from land. The extremely fine silk that spiders use for ballooning is known as gossamer.

In some cases, spiders may even use silk as a source of food.

Methods have been developed to collect silk from a spider by force.

Properties

Silk emitted by the silkworm consists of two main proteins, sericin and fibroin, fibroin being the structural centre of the silk, and serecin being the sticky material surrounding it. Fibroin is made up of the amino acids Gly-Ser-Gly-Ala-Gly-Ala and forms beta pleated sheets. Hydrogen bonds form between chains, and side chains form above and below the plane of the hydrogen bond network.

The high proportion (50%) of glycine, which is a small amino acid, allows tight packing and the fibers are strong and resistant to breaking. The tensile strength is due to the many interceded hydrogen bonds, and when stretched the force is applied to these numerous bonds and they do not break. Silk is resistant to most mineral acids, except for sulfuric acid, which dissolves it. It is yellowed by perspiration. Chlorine bleach will also destroy silk fabrics.

Mechanical Properties

Each spider and each type of silk has a set of mechanical properties optimised for their biological function.

Most silks, in particular dragline silk, have exceptional mechanical properties. They exhibit a unique combination of high tensile strength and extensibility (ductility). This enables a silk fiber to absorb a lot of energy before breaking (toughness, the area under a stress-strain curve).

A frequent mistake made in the mainstream media is to confuse strength and toughness when comparing silk to other materials. As shown below in detail, weight for weight, silk is stronger than steel, but not as strong as Kevlar. Silk is, however, tougher than both.

Strength

In detail a dragline silk's tensile strength is comparable to that of high-grade alloy steel (450 - 1970 MPa), and about half as strong as aramid filaments, such as Twaron or Kevlar (3000 MPa).

Density

Consisting of mainly protein, silks are about a sixth of the density of steel (1.31 g/cm^3). As a result, a strand long enough to circle the Earth would weigh less than 500 grams (18 oz). (Spider dragline silk has a tensile strength of roughly 1.3 GPa. The tensile strength listed for steel might be slightly higher—e.g. 1.65 GPa, but spider silk is a much less dense material, so that a given weight of spider silk is five times as strong as the same weight of steel.)

Energy Density

The energy density of dragline spider silk is $1.2x10^8 J/m^3$.

Extensibility

Silks are also extremely ductile, with some able to stretch up to five times their relaxed length without breaking.

Toughness

The combination of strength and ductility gives dragline silks a very high toughness (or work to fracture), which "equals that of commercial polyaramid (aromatic nylon) filaments, which themselves are benchmarks of modern polymer fiber technology".

Temperature

Whilst unlikely to be relevant in nature, dragline silks can hold their strength below "40°C (-40°F) and up to 220°C (428°F).

Supercontraction

When exposed to water, dragline silks undergo supercontraction, shrinking up to 50% in length and behaving like a weak rubber under tension. Many hypotheses have been suggested as to its use in nature, with the most popular being to automatically tension webs built in the night using the morning dew.

Highest-Performance

The toughest known spider silk is produced by the species Darwin's bark spider (*Caerostris darwini*): "The toughness of forcibly silked fibers averages 350 MJ/m^3, with some samples reaching 520 MJ/m^3. Thus, *C. darwini* silk is more than twice as tough as any previously described silk, and over 10 times tougher than Kevlar".

Macroscopic Structure Down to Protein Hierarchy

Silks, as well as many other biomaterials, have a hierarchical structure (e.g., cellulose, hair). The primary structure is its amino acid sequence, mainly consisting of highly repetitive glycine and alanine blocks, which is why silks are often referred to as a block co-polymer. On a secondary structure level, the short side chained alanine is mainly found in the crystalline domains (beta sheets) of the nanofibril, glycine is mostly found in the so-called amorphous matrix consisting of helical and beta turn structures. It is the interplay between the hard crystalline segments, and the strained elastic semi-amorphous regions, that gives spider silk its extraordinary properties. Various compounds other than protein are used to enhance the fiber's properties. Pyrrolidine has hygroscopic properties and helps to keep the thread moist. It occurs in especially high concentration in glue threads.

Potassium hydrogen phosphate releases protons in aqueous solution, resulting in a pH of about 4, making the silk acidic and thus protecting it from fungi and bacteria that would otherwise digest the protein. Potassium nitrate is believed to prevent the protein from denaturing in the acidic milieu.

This first very basic model of silk was introduced by Termonia in 1994 suggested crystallites embedded in an amorphous matrix interlinked with hydrogen bonds. This model has refined over the years: Semi-crystalline regions were found as well as a fibrillar skin core model suggested for spider silk, later visualized by AFM and TEM. Sizes of the nanofibrillar structure and the crystalline and semi-crystalline regions were revealed by neutron scattering.

Non-Protein Composition

Various compounds other than protein are found in spider silks, such as sugars, lipids, ions, and pigments that might affect the aggregation behaviour and act as a protection layer in the final fiber.

Biosynthesis

The production of silks, including spider silk, differs in an important respect from the production of most other fibrous biological materials: rather than being continuously grown as keratin in hair, cellulose in the cell walls of plants, or even the fibers formed from the compacted faecal matter of beetles, it is "spun" on demand from liquid silk precursor sometimes referred to as unspun silk dope, out of specialised glands.

The spinning process occurs when a fiber is pulled away from the body of a spider, be that by the spider's legs, by the spider's falling and using its own weight, or by any other method including being pulled by humans. The name "spinning" is misleading as no rotation of any component occurs, but the name comes from when it was thought that spiders produced their thread in a similar manner to the spinning wheels of old. In fact the process is a pultrusion—similar to extrusion, with the subtlety that the force is induced by pulling at the finished fiber rather than being squeezed out of a reservoir of some kind.

The unspun silk dope is pulled through silk glands, of which there may be both numerous duplicates and also different types on any one spider species.

Condensed Tannins

Condensed tannins (proanthocyanidins, polyflavonoid tannins, catechol-type tannins, pyrocatecollic type tannins, non-hydrolyzable tannins or flavolans) are polymers formed by the condensation of flavans. They do not contain sugar residues.

They are called proanthocyanidins as they yield anthocyanidins when depolymerized under oxidative conditions. Different types of condensed tannins exist, such as the procyanidins, propelargonidins, prodelphinidins, profisetinidins, proguibourtinidins or prorobinetidins, formed from flavonoids structures corresponding to the related anthocyanins. One particular type of condensed tannin, found in grape, are procyanidins, which are polymers of 2 to 50 (or more) flavan-3-ol units joined by carbon-carbon bonds. These are not susceptible to being cleaved by hydrolysis.

While many hydrolyzable tannins and most condensed tannins are water soluble, several tannins are also highly octanol soluble. Some large condensed tannins are insoluble. Differences in solubilities are likely to affect their biological functions.

Natural Occurrences

Tannins of tropical woods tend to be of a catechin nature rather than of the gallic type present in temperate woods.

Condensed tannins can be recovered from *Lithocarpus glaber* or can be found in *Prunus* sp. The bark of *Commiphora angolensis* contains condensed tannins.

Commercial sources of condensed tannins are plants such as quebracho wood (*Schinopsis lorentzii*), mimosa bark (*Acacia mollissima*), grapes seeds (*Vitis vinifera*), pine barks and spruce barks. Pycnogenol is a trademark for a French maritime pine bark extract.

Condensed tannins are formed in tannosomes, specialized organelles, in Tracheophytes i.e vascular plants.

Dietary Supplement

Pycnogenol is marketed as a dietary supplement with claims it can treat many conditions; however it is not effective for treating any chronic health disorder.

Analysis

Condensed tannins can be characterised by a number of modern techniques including depolymerisation, asymmetric flow field flow fractionation, small-angle X-ray scattering and MALDI-TOF mass spectrometry. Their interactions with proteins can be studied by isothermal titration calorimetry and this provides information on the affinity constant, enthalpy and stoichiometry in the tannin-protein complex.

Depolymerisation

Depolymerisation reactions are mainly analytical techniques but it is envisaged to use them as means to produce molecules for the chemical industry derived from waste products, such as bark from the wood industry or pomaces from the wine industry.

Depolymerisation is an indirect method of analysis allowing to gain informations such as average degree of polymerisation, percentage of galloylation, etc. The depolymerised sample can be injected on a mass spectrometre with an electrospray ionization source, only able to form ions with smaller molecules.

Oxidative Depolymerisation

The butanol–hydrochloric acid–iron assay (Porter assay) is a colorimetric assay. It is based on acid catalysed oxidative depolymerization of condensed tannins into corresponding anthocyanidins. The method has also been used for determination of bound condensed tannins, but has limitations. This reagent has recently been improved considerably by inclusion of acetone.

Non Oxidative Chemical Depolymerisation

The condensed tannins can nevertheless undergo acid-catalyzed cleavage in the presence of (an excess of) a nucleophile like phloroglucinol (reaction called phloroglucinolysis), benzyl mercaptan (reaction called thiolysis), thioglycolic acid (reaction called thioglycolysis) or cysteamine. These techniques are generally called depolymerisation and give informations such as average degree of polymerisation or percentage of galloylation.

These are SN1 reactions, a type of substitution reaction in organic chemistry, involving a carbocation intermediate under strongly acidic conditions in polar protic solvents like methanol. The reaction leads to the formation of free and derived monomers that can be further analyzed. The free monomers correspond to the terminal units of the condensed tannins chains. If thiolysis is done directly on plant material (rather than on purified tannins), it is, however, important to subtract naturally occurring free flavanol monomers from the concentration of terminal units that are released during depolymerisation.

Reactions are generally made in methanol, especially thiolysis, as benzyl mercaptan has a low solubility in water. They involve a moderate

(40 to 90°C) heating for a few minutes. Epimerisation may happen. Phloroglucinolysis can be used for instance for proanthocyanidins characterisation in wine or in the grape seed and skin tissues.

Thioglycolysis can be used to study proanthocyanidins or the oxidation of condensed tannins. It is also used for lignin quantitation. Reaction on condensed tannins from Douglas fir bark produces epicatechin and catechin thioglycolates.

Condensed tannins from *Lithocarpus glaber* leaves have been analysed through acid-catalyzed degradation in the presence of cysteamine.

DNA Sequence

DNA sequencing is the process of determining the precise order of nucleotides within a DNA molecule. It includes any method or technology that is used to determine the order of the four bases—adenine, guanine, cytosine, and thymine—in a strand of DNA. The advent of rapid DNA sequencing methods has greatly accelerated biological and medical research and discovery.

Knowledge of DNA sequences has become indispensable for basic biological research, and in numerous applied fields such as diagnostic, biotechnology, forensic biology, virology and biological systematics. The rapid speed of sequencing attained with modern DNA sequencing technology has been instrumental in the sequencing of complete DNA sequences, or genomes of numerous types and species of life, including the human genome and other complete DNA sequences of many animal, plant, and microbial species.

Use of Sequencing

DNA sequencing may be used to determine the sequence of individual genes, larger genetic regions (i.e. clusters of genes or operons), full chromosomes or entire genomes. Sequencing provides the order of individual nucleotides in DNA or RNA (commonly represented as A, C, G, T, and U) isolated from cells of animals, plants, bacteria, archaea, or virtually any other source of genetic information. This is useful for:

- Molecular biology – studying the genome itself, how proteins are made, what proteins are made, identifying new genes and associations with diseases and phenotypes, and identifying potential drug targets
- Evolutionary biology – studying how different organisms are related and how they evolved

- Metagenomics – Identifying species present in a body of water, sewage, dirt, debris filtred from the air, or swab samples of organisms. Helpful in ecology, epidemiology, microbiome research, and other fields.

Less-precise information is produced by non-sequencing techniques like DNA fingerprinting. This information may be easier to obtain and is useful for:

- Detect the presence of known genes for medical purposes
- Forensic identification
- Parental testing

History

Though the structure of DNA was established as a double helix in 1953, several decades would pass before fragments of DNA could be reliably analyzed for their sequence in the laboratory. RNA sequencing was one of the earliest forms of nucleotide sequencing. The major landmark of RNA sequencing is the sequence of the first complete gene and the complete genome of Bacteriophage MS2, identified and published by Walter Fiers and his coworkers at the University of Ghent (Ghent, Belgium), in 1972 and 1976.

The first method for determining DNA sequences involved a location-specific primer extension strategy established by Ray Wu at Cornell University in 1970. DNA polymerase catalysis and specific nucleotide labeling, both of which figure prominently in current sequencing schemes, were used to sequence the cohesive ends of lambda phage DNA Between 1970 and 1973, Wu, R Padmanabhan and colleagues demonstrated that this method can be employed to determine any DNA sequence using synthetic location-specific primers. Frederick Sanger then adopted this primer-extension strategy to develop more rapid DNA sequencing methods at the MRC Centre, Cambridge, UK and published a method for "DNA sequencing with chain-terminating inhibitors" in 1977. Walter Gilbert and Allan Maxam at Harvard also developed sequencing methods, including one for "DNA sequencing by chemical degradation". In 1973, Gilbert and Maxam reported the sequence of 24 basepairs using a method known as wandering-spot analysis. Advancements in sequencing were aided by the concurrent development of recombinant DNA technology, allowing DNA samples to be isolated from sources other than viruses.

The first full DNA genome to be sequenced was that of bacteriophage φX174 in 1977. Medical Research Council scientists deciphered the complete DNA sequence of the Epstein-Barr virus in

1984, finding it to be 170 thousand base-pairs long. A non-radioactive method for transferring the DNA molecules of sequencing reaction mixtures onto an immobilizing matrix during electrophoresis was developed by Pohl and co-workers in the early 80's. Followed by the commercialization of the DNA sequencer "Direct-Blotting-Electrophoresis-System GATC 1500" by GATC Biotech, which was intensively used in the framework of the EU genome-sequencing programme, the complete DNA sequence of the yeast *Saccharomyces cerevisiae* chromosome II. Leroy E. Hood's laboratory at the California Institute of Technology announced the first semi-automated DNA sequencing machine in 1986.

This was followed by Applied Biosystems' marketing of the first fully automated sequencing machine, the ABI 370, in 1987 and by Dupont's Genesis 2000 which used a novel fluorescent labeling technique enabling all four dideoxynucleotides to be identified in a single lane. By 1990, the U.S. National Institutes of Health (NIH) had begun large-scale sequencing trials on *Mycoplasma capricolum*, *Escherichia coli*, *Caenorhabditis elegans*, and *Saccharomyces cerevisiae* at a cost of US$0.75 per base. Meanwhile, sequencing of human cDNA sequences called expressed sequence tags began in Craig Venter's lab, an attempt to capture the coding fraction of the human genome. In 1995, Venter, Hamilton Smith, and colleagues at The Institute for Genomic Research (TIGR) published the first complete genome of a free-living organism, the bacterium *Haemophilus influenzae*. The circular chromosome contains 1,830,137 bases and its publication in the journal Science marked the first published use of whole-genome shotgun sequencing, eliminating the need for initial mapping efforts. By 2001, shotgun sequencing methods had been used to produce a draft sequence of the human genome.

Several new methods for DNA sequencing were developed in the mid to late 1990s. These techniques comprise the first of the "next-generation" sequencing methods. In 1996, Pɛl Nyrén and his student Mostafa Ronaghi at the Royal Institute of Technology in Stockholm published their method of pyrosequencing. A year later, Pascal Mayer and Laurent Farinelli submitted patents to the World Intellectual Property Organization describing DNA colony sequencing. Lynx Therapeutics published and marketed "Massively parallel signature sequencing", or MPSS, in 2000. This method incorporated a parallelized, adapter/ligation-mediated, bead-based sequencing technology and served as the first commercially available "next-generation" sequencing method, though no DNA sequencers were sold to independent laboratories. In 2004, 454 Life Sciences marketed a parallelized version of pyrosequencing. The first version of their machine reduced sequencing costs 6-fold compared to automated Sanger sequencing, and was the second of the new generation of sequencing technologies, after MPSS.

The large quantities of data produced by DNA sequencing have also required development of new methods and programs for sequence analysis. Phil Green and Brent Ewing of the University of Washington described their phred quality score for sequencer data analysis in 1998.

Basic Methods

Maxam-Gilbert Sequencing

Allan Maxam and Walter Gilbert published a DNA sequencing method in 1977 based on chemical modification of DNA and subsequent cleavage at specific bases. Also known as chemical sequencing, this method allowed purified samples of double-stranded DNA to be used without further cloning. This method's use of radioactive labeling and its technical complexity discouraged extensive use after refinements in the Sanger methods had been made.

Maxam-Gilbert sequencing requires radioactive labeling at one 5' end of the DNA and purification of the DNA fragment to be sequenced. Chemical treatment then generates breaks at a small proportion of one or two of the four nucleotide bases in each of four reactions (G, A+G, C, C+T). The concentration of the modifying chemicals is controlled to introduce on average one modification per DNA molecule. Thus a series of labeled fragments is generated, from the radiolabeled end to the first "cut" site in each molecule. The fragments in the four reactions are electrophoresed side by side in denaturing acrylamide gels for size separation. To visualize the fragments, the gel is exposed to X-ray film for autoradiography, yielding a series of dark bands each corresponding to a radiolabeled DNA fragment, from which the sequence may be inferred.

Chain-Termination Methods

The chain-termination method developed by Frederick Sanger and coworkers in 1977 soon became the method of choice, owing to its relative ease and reliability. When invented, the chain-terminator method used fewer toxic chemicals and lower amounts of radioactivity than the Maxam and Gilbert method. Because of its comparative ease, the Sanger method was soon automated and was the method used in the first generation of DNA sequencers.

Sanger sequencing is the method which prevailed from the 80's until the mid-2000s. Over that period, great advances were made in the technique, such as fluorescent labelling, capillary electrophoresis, and general automation. These developments allowed much more efficient sequencing, leading to lower costs. The Sanger method, in mass production form, is the technology which produced the first human genome in 2001, ushering in the age of genomics. However, later in the decade, radically different approaches reached the market, bringing the cost per genome down from $100 million in 2001 to $10,000 in 2011.

Advanced Methods and de novo Sequencing

Large-scale sequencing often aims at sequencing very long DNA pieces, such as whole chromosomes, although large-scale sequencing can also be used to generate very large numbers of short sequences, such as found in phage display. For longer targets such as chromosomes, common approaches consist of cutting (with restriction enzymes) or shearing (with mechanical forces) large DNA fragments into shorter DNA fragments. The fragmented DNA may then be cloned into a DNA vector and amplified in a bacterial host such as *Escherichia coli*. Short DNA fragments purified from individual bacterial colonies are individually sequenced and assembled electronically into one long, contiguous sequence. Studies have shown that adding a size selection step to collect DNA fragments of uniform size can improve sequencing efficiency and accuracy of the genome assembly. In these studies, automated sizing has proven to be more reproducible and precise than manual gel sizing.

The term "*de novo* sequencing" specifically refers to methods used to determine the sequence of DNA with no previously known sequence. *De novo* translates from Latin as "from the beginning". Gaps in the assembled sequence may be filled by primer walking. The different strategies have different tradeoffs in speed and accuracy; shotgun methods are often used for sequencing large genomes, but its assembly is complex and difficult, particularly with sequence repeats often causing gaps in genome assembly.

Most sequencing approaches use an *in vitro* cloning step to amplify individual DNA molecules, because their molecular detection methods are not sensitive enough for single molecule sequencing. Emulsion PCR isolates individual DNA molecules along with primer-coated beads in aqueous droplets within an oil phase. A polymerase chain reaction (PCR) then coats each bead with clonal copies of the DNA molecule followed by immobilization for later sequencing. Emulsion PCR is used in the methods developed by Marguilis et al. (commercialized by 454 Life Sciences), Shendure and Porreca et al. (also known as "Polony sequencing") and SOLiD sequencing, (developed by Agencourt, later Applied Biosystems, now Life Technologies).

Shotgun Sequencing

Shotgun sequencing is a sequencing method designed for analysis of DNA sequences longer than 1000 base pairs, up to and including entire chromosomes. This method requires the target DNA to be broken into random fragments. After sequencing individual fragments, the sequences can be reassembled on the basis of their overlapping regions.

Bridge PCR

Another method for *in vitro* clonal amplification is bridge PCR, in which fragments are amplified upon primers attached to a solid surface and form "DNA colonies" or "DNA clusters". This method is used in the Illumina Genome Analyzer sequencers. Single-molecule methods, such as that developed by Stephen Quake's laboratory (later commercialized by Helicos) are an exception: they use bright fluorophores and laser excitation to detect base addition events from individual DNA molecules fixed to a surface, eliminating the need for molecular amplification.

Next-Generation Methods

Next-generation sequencing applies to genome sequencing, genome resequencing, transcriptome profiling (RNA-Seq), DNA-protein interactions (ChIP-sequencing), and epigenome characterization. Resequencing is necessary, because the genome of a single individual of a species will not indicate all of the genome variations among other individuals of the same species.

The high demand for low-cost sequencing has driven the development of high-throughput sequencing (or next-generation sequencing) technologies that parallelize the sequencing process, producing thousands or millions of sequences concurrently. High-throughput sequencing technologies are intended to lower the cost of DNA sequencing beyond what is possible with standard dye-terminator methods. In ultra-high-throughput sequencing as many as 500,000 sequencing-by-synthesis operations may be run in parallel.

Methods in Development

DNA sequencing methods currently under development include labeling the DNA polymerase, reading the sequence as a DNA strand transits through nanopores, and microscopy-based techniques, such as atomic force microscopy or transmission electron microscopy that are used to identify the positions of individual nucleotides within long DNA fragments (>5,000 bp) by nucleotide labeling with heavier elements (e.g., halogens) for visual detection and recording.

Third generation technologies aim to increase throughput and decrease the time to result and cost by eliminating the need for excessive reagents and harnessing the processivity of DNA polymerase.

Nanopore DNA Sequencing

This method is based on the readout of electrical signals occurring at nucleotides passing by alpha-hemolysin pores covalently bound with cyclodextrin. The DNA passing through the nanopore changes its ion current. This change is dependent on the shape, size and length of the DNA sequence. Each type of the nucleotide blocks the ion flow through the pore for a different period of time. The method has a potential of development as it does not require modified nucleotides, however single nucleotide resolution is not yet available.

Two main areas of nanopore sequencing in development are solid state nanopore sequencing, and protein based nanopore sequencing. Protein nanopore sequencing utilizes membrane protein complexes ""-Hemolysin and MspA (Mycobacterium Smegmatis Porin A), which show great promise given their ability to distinguish between individual and groups of nucleotides. Whereas, solid-state nanopore sequencing utilizes synthetic materials such as silicon nitride and aluminum oxide and it is preferred for its superior mechanical ability and thermal and chemical stability. The fabrication method is essential for this type of sequencing given that the nanopore array can contain hundreds of pores with diameters smaller than eight nanometres.

The concept originated from the idea that single stranded DNA or RNA molecules can be electrophoretically driven in a strict linear sequence through a biological pore that can be less than eight nanometres, and can be detected given that the molecules release an ionic current while moving through the pore.

The pore contains a detection region capable of recognizing different bases, with each base generating various time specific signals corresponding to the sequence of bases as they cross the pore which are then evaluated. When implementing this process it is important to note that precise control over the DNA transport through the pore is crucial for success. Various enzymes such as exonucleases and polymerases have been used to moderate this process by positioning them near the pore's entrance.

Tunnelling Currents DNA Sequencing

Another approach uses measurements of the electrical tunnelling currents across single-strand DNA as it moves through a channel. Depending on its electronic structure each base affects the tunnelling current differently, allowing differentiation between different bases.

The use of tunnelling currents has the potential to sequence orders of magnitude faster than ionic current methods and the sequencing of several DNA oligomers and micro-RNA has already been achieved.

Sequencing by Hybridization

Sequencing by hybridization is a non-enzymatic method that uses a DNA microarray. A single pool of DNA whose sequence is to be determined is fluorescently labeled and hybridized to an array containing known sequences. Strong hybridization signals from a given spot on the array identifies its sequence in the DNA being sequenced.

This method of sequencing utilizes binding characteristics of a library of short single stranded DNA molecules (oligonucleotides) also called DNA probes to reconstruct a target DNA sequence. Non-specific hybrids are removed by washing and the target DNA is eluted. Hybrids are re-arranged such that the DNA sequence can be reconstructed. The benefit of this sequencing type is its ability to capture a large number of targets with a homogenous coverage. Although a large number of chemicals and starting DNA is usually required. But, with the advent of solution-based hybridization much less equipment and chemicals are necessary.

Sequencing with Mass Spectrometry

Mass spectrometry may be used to determine DNA sequences. Matrix-assisted laser desorption ionization time-of-flight mass spectrometry, or MALDI-TOF MS, has specifically been investigated as an alternative method to gel electrophoresis for visualizing DNA fragments. With this method, DNA fragments generated by chain-termination sequencing reactions are compared by mass rather than by size. The mass of each nucleotide is different from the others and this difference is detectable by mass spectrometry. Single-nucleotide mutations in a fragment can be more easily detected with MS than by gel electrophoresis alone. MALDI-TOF MS can more easily detect differences between RNA fragments, so researchers may indirectly sequence DNA with MS-based methods by converting it to RNA first.

The higher resolution of DNA fragments permitted by MS-based methods is of special interest to researchers in forensic science, as they may wish to find single-nucleotide polymorphisms in human DNA samples to identify individuals. These samples may be highly degraded so forensic researchers often prefer mitochondrial DNA for its higher stability and applications for lineage studies. MS-based sequencing methods have been used to compare the sequences of human mitochondrial DNA from samples in a Federal Bureau of Investigation database and from bones found in mass graves of World War I soldiers.

Early chain-termination and TOF MS methods demonstrated read lengths of up to 100 base pairs. Researchers have been unable to exceed this average read size; like chain-termination sequencing alone, MS-based DNA sequencing may not be suitable for large *de novo* sequencing projects. Even so, a recent study did use the short sequence reads and mass spectroscopy to compare single-nucleotide polymorphisms in pathogenic *Streptococcus* strains.

Microfluidic Sanger Sequencing

In microfluidic Sanger sequencing the entire thermocycling amplification of DNA fragments as well as their separation by electrophoresis is done on a single glass wafer (approximately 10 cm in diameter) thus reducing the reagent usage as well as cost. In some instances researchers have shown that they can increase the throughput of conventional sequencing through the use of microchips. Research will still need to be done in order to make this use of technology effective.

Microscopy-Based Techniques

This approach directly visualizes the sequence of DNA molecules using electron microscopy. The first identification of DNA base pairs within intact DNA molecules by enzymatically incorporating modified bases, which contain atoms of increased atomic number, direct visualization and identification of individually labeled bases within a synthetic 3,272 base-pair DNA molecule and a 7,249 base-pair viral genome has been demonstrated.

RNAP Sequencing

This method is based on use of RNA polymerase (RNAP), which is attached to a polystyrene bead. One end of DNA to be sequenced is attached to another bead, with both beads being placed in optical traps. RNAP motion during transcription brings the beads in closer and their relative distance changes, which can then be recorded at a single nucleotide resolution. The sequence is deduced based on the four readouts with lowered concentrations of each of the four nucleotide types, similarly to the Sanger method.

RNA polymerase is attached to one end of a polystyrene bead and the other end is attached to the distal end of a DNA fragment. Each bead is then stuck in to an optical trap that levitates the beads. The interactions between the RNAP and the DNA result in a change in the length of the DNA between the two beads. This change is the measured with precision resulting in a single base resolution on a single DNA molecule. This is then repeated four times where each time there is a lower concentration of one of the four nucleotides, this shares some similarity with the primers used in the Sanger Sequencing method. A comparison is made between regions and sequence information is deduced by comparing the known sequence regions to the unknown sequence regions.

In Vitro *Virus High-Throughput Sequencing*

A method has been developed to analyze full sets of protein interactions using a combination of 454 pyrosequencing and an *in vitro* virus mRNA display method. Specifically, this method covalently links proteins of interest to the mRNAs encoding them, then detects the mRNA pieces using reverse transcription PCRs. The mRNA may then be amplified and sequenced. The combined method was titled IVV-HiTSeq and can be performed under cell-free conditions, though its results may not be representative of *in vivo* conditions.

Development Initiatives

In October 2006, the X Prize Foundation established an initiative to promote the development of full genome sequencing technologies, called the Archon X Prize, intending to award $10 million to "the first Team that can build a device and use it to sequence 100 human genomes within 10 days or less, with an accuracy of no more than one error in every 100,000 bases sequenced, with sequences accurately covering at least 98% of the genome, and at a recurring cost of no more than $10,000 (US) per genome."

Each year the National Human Genome Research Institute, or NHGRI, promotes grants for new research and developments in genomics. 2010 grants and 2011 candidates include continuing work in microfluidic, polony and base-heavy sequencing methodologies.

Computational Challenges

The sequencing technologies described here produce raw data that needs to be assembled into longer sequences such as complete genomes (sequence assembly). There are many computational challenges to achieve this, such as the evaluation of the raw sequence data which is done by programs and algorithms such as Phred and Phrap. Other challenges have to deal with repetitive sequences that often prevent complete genome assemblies because they occur in many places of the genome. As a consequence, many sequences may not be assigned to particular chromosomes. The production of raw sequence data is only the beginning of its detailed bioinformatical analysis. Yet new methods for sequencing and correcting sequencing errors were developed. [1]

Read Trimming

Sometimes, the raw reads produced by the sequencer are correct and precise only in a fraction of their length. Using the entire read may introduce artifacts in the downstream analyses like genome assembly, snp calling, or gene expression estimation. Two classes of trimming programs have been introduced, based on the window-based or the running-sum classes of algorithms.

Melanin

Melanin is a broad term for a group of natural pigments found in most organisms (arachnids are one of the few groups in which it has not been detected). Melanin is produced by the oxidation of the amino acid tyrosine, followed by polymerization. The pigment is produced in a specialized group of cells known as melanocytes.

There are three basic types of melanin: eumelanin, pheomelanin, and neuromelanin. The most common type is eumelanin, and is produced in 'black' and 'brown' subtypes. Pheomelanin is a cysteine-containing red-brown polymer of benzothiazine units largely responsible for red hair and freckles. Neuromelanin is found in the brain, though its function remains obscure.

In the skin, melanogenesis occurs after exposure to UV radiation, causing the skin to visibly tan. Melanin is an effective absorber of light; the pigment is able to dissipate over 99.9% of absorbed UV radiation. Because of this property, melanin is thought to protect skin cells from UVB radiation damage, reducing the risk of cancer. Furthermore, though exposure to UV radiation is associated with increased risk of malignant melanoma, a cancer of the melanocytes, studies have shown a lower incidence for skin cancer in individuals with more concentrated melanin, i.e. darker skin tone. Nonetheless, the relationship between skin pigmentation and photoprotection is still being clarified.

Humans

In humans, melanin is the primary determinant of skin colour. It is also found in hair, the pigmented tissue underlying the iris of the eye, and the stria vascularis of the inner ear. In the brain, tissues with melanin include the medulla and pigment-bearing neurons within areas of the brainstem, such as the locus coeruleus and the substantia nigra. It also occurs in the zona reticularis of the adrenal gland.

The melanin in the skin is produced by melanocytes, which are found in the basal layer of the epidermis. Although, in general, human beings possess a similar concentration of melanocytes in their skin, the melanocytes in some individuals and ethnic groups more frequently or less frequently express and humans have very little or no melanin synthesis in their bodies, a condition known as albinism.

Because melanin is an aggregate of smaller component molecules, there are many different types of melanin with differing proportions and bonding patterns of these component molecules. Both pheomelanin and eumelanin are found in human skin and hair, but eumelanin is the most abundant melanin in humans, as well as the form most likely to be deficient in albinism.

Eumelanin

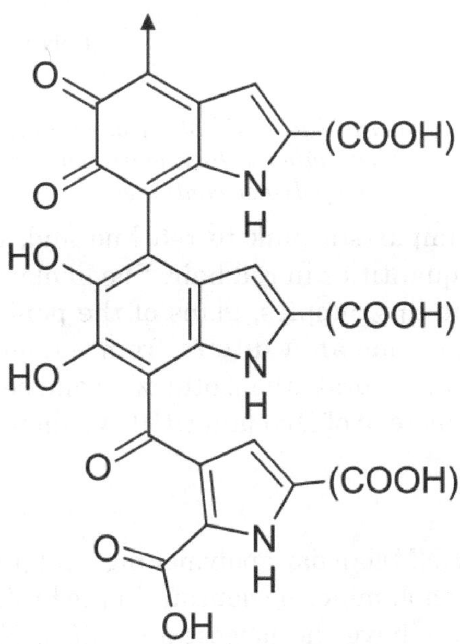

Figure: *Part of the structural formula of eumelanin. "(COOH)" can be COOH or H, or (more rarely) other substituents. The arrow denotes where the polymer continues.*

Eumelanin polymers have long been thought to comprise numerous cross-linked 5,6-dihydroxyindole (DHI) and 5,6-dihydroxyindole-2-

carboxylic acid (DHICA) polymers. Two types are recognized: black and brown. Black melanin is darker than brown. A small amount of black eumelanin in the absence of other pigments causes grey hair. A small amount of brown eumelanin in the absence of other pigments causes yellow (blond) colour hair.

Pheomelanin

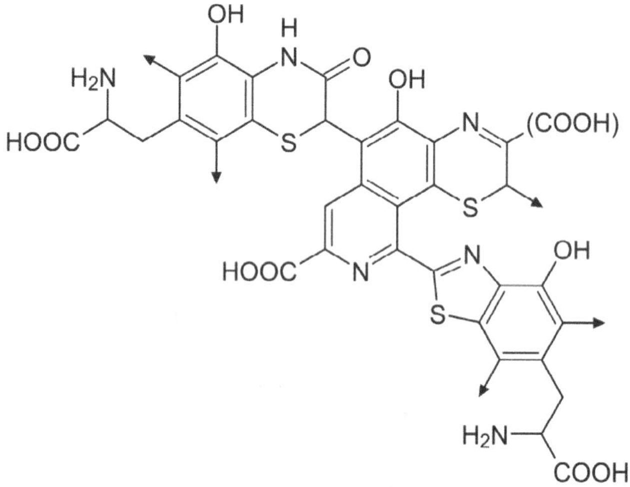

Figure: *Part of the structural formula of pheomelanin. "(COOH)" can be COOH or H, or (more rarely) other substituents. The arrows denote where the polymer continues.*

Pheomelanin imparts a pink to red hue and, thus, is found in particularly large quantities in red hair. Pheomelanin is particularly concentrated in the lips, nipples, glans of the penis, and vagina. In chemical terms, pheomelanin differs from eumelanin in that its oligomer structure incorporates benzothiazine and benzothiazole units that are produced, instead of DHI and DHICA, when the amino acid L-cysteine is present.

Neuromelanin

Neuromelanin (NM) is a dark polymer pigment produced in specific populations of catecholaminergic neurons in the brain, or in the upper vocal cords. Humans have the largest amount of NM, while it is in lesser amount in other non-human primates, and totally absent in other species. However, the biological function remains unknown, although human NM has been shown to efficiently bind transition metals such as iron, as well as other potentially toxic molecules. Therefore, it may play crucial roles in apoptosis and the related Parkinson's disease.

Other Organisms

Melanins have very diverse roles and functions in various organisms. A form of melanin makes up the ink used by many cephalopods as a defense mechanism against predators. Melanins also protect microorganisms, such as bacteria and fungi, against stresses that involve cell damage such as UV radiation from the sun and reactive oxygen species. Melanin also protects against damage from high temperatures, chemical stresses (such as heavy metals and oxidizing agents), and biochemical threats (such as host defenses against invading microbes). Therefore, in many pathogenic microbes (for example, in *Cryptococcus neoformans*, a fungus) melanins appear to play important roles in virulence and pathogenicity by protecting the microbe against immune responses of its host. In invertebrates, a major aspect of the innate immune defense system against invading pathogens involves melanin. Within minutes after infection, the microbe is encapsulated within melanin (melanization), and the generation of free radical byproducts during the formation of this capsule is thought to aid in killing them. Some types of fungi, called radiotrophic fungi, appear to be able to use melanin as a photosynthetic pigment that enables them to capture gamma rays and harness this energy for growth.

The black feathers of birds owe their colour to melanin; they are less readily degraded by bacteria than white feathers, or those containing other pigments such as carotenes. In a bird's eye, a specialized organ, rich in blood vessels, the pecten oculi, is also extremely rich in melanin, which has been considered to have role in absorption of light falling on optic disc and using it to warm up the eye. This, in turn may stimulate release of nutrients from pecten oculi to retina, via vitreous humor; it is plausible as bird retina is devoid of its own blood vessels. In pigment epithelium of retina, presence of high amounts of melanin granules may also minimize back-scatter of image light on the retina.

In some mice, melanin is used slightly differently. For instance, in Agouti mice, the hair appears brown because of alternation between black eumelanin production and a yellow variety of pheomelanin. The hairs are actually banded black and yellow, and the net effect is the brown colour of most mice. Some genetic irregularities can produce either fully black or fully yellow mice.

Microscopic Appearance

Melanin is brown, non-refractile, and finely granular with individual granules having a diameter of less than 800 nanometres.

This differentiates melanin from common blood breakdown pigments, which are larger, chunky, and refractile, and range in colour from green to yellow or red-brown. In heavily pigmented lesions, dense aggregates of melanin can obscure histologic detail. A dilute solution of potassium permanganate is an effective melanin bleach.

Genetic Disorders and Disease States

Melanin deficiency has been connected for some time with various genetic abnormalities and disease states.

There are approximately nine different types of oculocutaneous albinism, which is mostly an autosomal recessive disorder. Certain ethnicities have higher incidences of different forms. For example, the most common type, called oculocutaneous albinism type 2 (OCA2), is especially frequent among people of black African descent. It is an autosomal recessive disorder characterized by a congenital reduction or absence of melanin pigment in the skin, hair, and eyes. The estimated frequency of OCA2 among African-Americans is 1 in 10,000, which contrasts with a frequency of 1 in 36,000 in white Americans. In some African nations, the frequency of the disorder is even higher, ranging from 1 in 2,000 to 1 in 5,000. Another form of Albinism, the "yellow oculocutaneous albinism", appears to be more prevalent among the Amish, who are of primarily Swiss and German ancestry. People with this IB variant of the disorder commonly have white hair and skin at birth, but rapidly develop normal skin pigmentation in infancy.

Ocular albinism affects not only eye pigmentation but visual acuity, as well. People with albinism typically test poorly, within the 20/60 to 20/400 range. In addition, two forms of albinism, with approximately 1 in 2700 most prevalent among people of Puerto Rican origin, are associated with mortality beyond melanoma-related deaths.

Mortality also is increased in patients with Hermansky-Pudlak syndrome and Chediak-Higashi syndrome. Patients with Hermansky-Pudlak syndrome have a bleeding diathesis secondary to platelet dysfunction and also experience restrictive lung disease (pulmonary fibrosis), inflammatory bowel disease, cardiomyopathy, and renal disease. Patients with Chediak-Higashi syndrome are susceptible to infection and also can develop lymphofollicular malignancy.

The role that melanin deficiency plays in such disorders remains under study. The connection between albinism and deafness is well known, though poorly understood. In his 1859 treatise *On the Origin of Species*, Charles Darwin observed that "cats which are entirely white and have blue eyes are generally deaf". In humans, hypopigmentation and deafness occur together in the rare Waardenburg's syndrome, predominantly observed among the Hopi in North America. The incidence of albinism in Hopi Indians has been estimated as approximately 1 in 200 individuals. It is interesting to note that similar patterns of albinism and deafness have been found in other mammals, including dogs and rodents. However, a lack of melanin *per se* does not appear to be directly responsible for deafness associated with hypopigmentation, as most individuals lacking the enzymes required to synthesize melanin have normal auditory function. Instead the absence of melanocytes in the stria vascularis of the inner ear results in cochlear impairment, though why this is, is not fully understood.

In Parkinson's disease, a disorder that affects neuromotor functioning, there is decreased neuromelanin in the substantia nigra and locus coeruleus as consequence of specific dropping out of dopaminergic and noradrenergic pigmented neurons. This results in diminished dopamine and norepinephrine synthesis. While no correlation between race and the level of neuromelanin in the substantia nigra has been reported, the significantly lower incidence of Parkinson's in blacks than in whites has "prompt[ed] some to suggest that cutaneous melanin might somehow serve to protect the neuromelanin in substantia nigra from external toxins.".

In addition to melanin deficiency, the molecular weight of the melanin polymer may be decreased by various factors such as oxidative stress, exposure to light, perturbation in its association with melanosomal matrix proteins, changes in pH, or in local concentrations of metal ions. A decreased molecular weight or a decrease in the degree of polymerization of ocular melanin has been proposed to turn the normally anti-oxidant polymer into a pro-oxidant. In its pro-oxidant state, melanin has been suggested to be involved in the causation and progression of macular degeneration and melanoma. Rasagiline, an important monotherapy drug in Parkinson's disease, has melanin binding properties, and melanoma tumor reducing properties.

Higher eumelanin levels also can be a disadvantage, however, beyond a higher disposition toward vitamin D deficiency. Dark skin is a complicating factor in the laser removal of port-wine stains. Effective in treating white skin, in general, lasers are less successful in removing port-wine stains in people of Asian or African descent. Higher concentrations of melanin in darker-skinned individuals simply diffuse and absorb the laser radiation, inhibiting light absorption by the targeted tissue. In similar manner, melanin can complicate laser treatment of other dermatological conditions in people with darker skin.

Freckles and moles are formed where there is a localized concentration of melanin in the skin. They are highly associated with pale skin.

Nicotine has an affinity for melanin-containing tissues because of its precursor function in melanin synthesis or its irreversible binding of melanin. This has been suggested to underlie the increased nicotine dependence and lower smoking cessation rates in darker pigmented individuals.

Human Adaptation

Melanocytes insert granules of melanin into specialized cellular vesicles called melanosomes. These are then transferred into the other skin cells of the human epidermis. The melanosomes in each recipient cell accumulate atop the cell nucleus, where they protect the nuclear DNA from mutations caused by the ionizing radiation of the sun's ultraviolet rays. In general, people whose ancestors lived for long periods in the regions of the globe near the equator have larger quantities of eumelanin in their skins. This makes their skins brown or black and protects them against high levels of exposure to the sun, which more frequently result in melanomas in lighter-skinned people.

With humans, exposure to sunlight stimulates the skin to produce vitamin D. Because high levels of cutaneous melanin act as a natural sun screen, dark skin can be a risk factor for vitamin D deficiency in regions of the Earth known as cool temperate zones, i.e., above 36 degrees latitude in the Northern hemisphere and below 36 degrees in the Southern hemisphere. As a result of this, health authorities in Canada and the USA have issued recommendations for people with darker complexions (including people of southern European descent) to consume between 1000-2000 IU (International Units) of vitamin D, daily, autumn through spring.

The most recent scientific evidence indicates that all humans originated in Africa, then populated the rest of the world through successive radiations. It seems likely that the first modern humans had relatively large numbers of eumelanin-producing melanocytes. In accordance, they had darker skin as with the indigenous people of Africa today. As some of these original peoples migrated and settled in areas of Asia and Europe, the selective pressure for eumelanin production decreased in climates where radiation from the sun was less intense. Of the two common gene variants known to be associated with pale human skin, *Mc1r* does not appear to have undergone positive selection, while *SLC24A5* has.

As with peoples having migrated northward, those with light skin migrating toward the equator acclimatize to the much stronger solar radiation. Most people's skin darkens when exposed to UV light, giving them more protection when it is needed. This is the physiological purpose of sun tanning. Dark-skinned people, who produce more skin-protecting eumelanin, have a greater protection against sunburn and the development of melanoma, a potentially deadly form of skin cancer, as well as other health problems related to exposure to strong solar radiation, including the photodegradation of certain vitamins such as riboflavins, carotenoids, tocopherol, and folate.

Melanin in the eyes, in the iris and choroid, helps protect them from ultraviolet and high-frequency visible light; people with gray, blue, and green eyes are more at risk for sun-related eye problems. Further, the ocular lens yellows with age, providing added protection. However, the lens also becomes more rigid with age, losing most of its accommodation — the ability to change shape to focus from far to near — a detriment due probably to protein crosslinking caused by UV exposure.

Recent research by J.D. Simon *et al.* suggests that melanin may serve a protective role other than photoprotection. Melanin is able to effectively ligate metal ions through its carboxylate and phenolic hydroxyl groups, in many cases much more efficiently than the powerful chelating ligand ethylenediaminetetraacetate (EDTA). Thus, it may serve to sequester potentially toxic metal ions, protecting the rest of the cell. This hypothesis is supported by the fact that the loss of neuromelanin observed in Parkinson's disease is accompanied by an increase in iron levels in the brain.

Physical Properties and Technological Applications

Evidence exists in support of a highly cross-linked heteropolymer bound covalently to matrix scaffolding melanoproteins. It has been proposed that the ability of melanin to act as an antioxidant is directly proportional to its degree of polymerization or molecular weight. Suboptimal conditions for the effective polymerization of melanin monomers may lead to formation of lower-molecular-weight, pro-oxidant melanin that has been implicated in the causation and progression of macular degeneration and melanoma. Signaling pathways that upregulate melanization in the retinal pigment epithelium (RPE) also may be implicated in the downregulation of rod outer segment phagocytosis by the RPE. This phenomenon has been attributed in part to foveal sparing in macular degeneration.

Phosphoramidite

A phosphoramidite $(RO)_2PNR_2$ is a monoamide of a phosphite diester. The key feature of phosphoramidites is their markedly high reactivity towards nucleophiles catalyzed by weak acids *e.c.*, triethylammonium chloride or 1*H*-tetrazole. In these reactions, the incoming nucleophile replaces the NR_2 moiety.

Nucleoside Phosphoramidites

Phosphoramidites derived from protected nucleosides are referred to as nucleoside phosphoramidites and are widely used in chemical synthesis of DNA, RNA, and other nucleic acids and their analogs.

As Ligands

Certain phosphoramidites are also used as monodentate chiral ligands in asymmetric synthesis. A large group of such ligands is derived from the chiral diol BINOL and can be synthesis by reaction of BINOL with phosphorus trichloride to the chlorophosphite and then reaction with simple secondary amines. This type of ligand was first used in 1996 in an asymmetric copper-catalysed addition of dialkylzincs to enones

Small Molecules

In molecular biology and pharmacology, a small molecule is a low molecular weight (<900 daltons[1]) organic compound that may help regulate a biological process, with a size on the order of 10^{-9} m. Most drugs are small molecules.

The upper molecular weight limit for a small molecule is approximately 900 daltons, which allows for the possibility to rapidly diffuse across cell membranes so that they can reach intracellular sites of action. In addition, this molecular weight cutoff is a necessary but insufficient condition for oral bioavailability. Finally, a lower molecular weight cutoff of 500 daltons (as part of the "rule of five") has been recommended for small molecule drug development candidates based on the observation that clinical attrition rates are significantly reduced if the molecular weight is kept below this 500 dalton limit.

Pharmacology usually restricts the term to a molecule that binds to a specific biopolymer—such as protein or nucleic acid—and acts as an effector, altering the activity or function of the biopolymer. Small molecules can have a variety of biological functions, serving as cell signaling molecules, as drugs in medicine, as pesticides in farming, and in many other roles. These compounds can be natural (such as secondary metabolites) or artificial (such as antiviral drugs); they may have a beneficial effect against a disease (such as drugs) or may be detrimental (such as teratogens and carcinogens). Biopolymers such as nucleic acids and proteins, and polysaccharides (such as starch or cellulose) are not small molecules—though their constituent monomers—ribo- or deoxyribonucleotides, amino acids, and monosaccharides, respectively—are often considered small molecules. Very small oligomers are also usually considered small molecules, such as dinucleotides, peptides such as the antioxidant glutathione, and disaccharides such as sucrose.

Small molecules may also be used as research tools to probe biological function as well as leads in the development of new therapeutic agents. Some can inhibit a specific function of a multifunctional protein or disrupt protein–protein interactions.

Drugs

Most drugs are small molecules, although some drugs can be proteins (e.g., insulin and other biologic medical products). Many proteins are degraded if administered orally and most often cannot cross cell membranes. Small molecules are more likely to be absorbed, although some of them are only absorbed after oral administration if given as prodrugs. One advantage small molecule drugs (SMDs) have over "large molecule" biologics is that many SMDs can be taken orally whereas biologics generally require injection or another parenteral administration.

Secondary Metabolites

A wide variety of organisms including bacteria, fungi, and plants, produce small molecule secondary metabolites also known as natural products, which play a role in cell signalling, pigmentation and in defense against predation. Secondary metabolites are a rich source of biologically active compounds and hence are often used as research tools and leads for drug discovery. Examples of secondary metabolites include:

- Alkaloids
- Glycosides
- Lipids
- Nonribosomal peptides, such as actinomycin-D
- Phenazines
- Natural phenols (including flavonoids)
- Polyketide

- Terpenes, including steroids
- Tetrapyrroles.

Research Tools

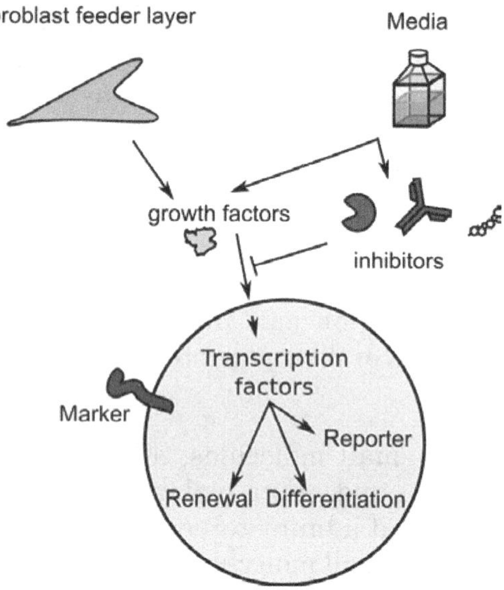

Figure: *Cell culture example of a small molecule as a tool instead of a protein. in cell culture to obtain a pancreatic lineage from mesodermal stem cells the retinoic acid signalling pathway must be activated while the sonic hedgehog pathway inhibited, which can be done by adding to the media anti-shh antibodies, Hedgehog interacting protein or cyclopamine, the first two are protein and the last a small molecule.*

Enzymes and receptors are often activated or inhibited by endogenous protein, but can be also inhibited by endogenous or exogenous small molecule inhibitors or activators which can bind to the active site or on the allosteric site. An example is the teratogen and carcinogen phorbol 12-myristate 13-acetate, which is a plant terpene that activates protein kinase C, which promotes cancer, making it a useful investigative tool. There is also interest in creating small molecule artificial transcription factors to regulate gene expression, examples include wrenchnolol (a wrench shaped molecule).

Binding of ligand can be characterised using a variety of analytical techniques such as surface plasmon resonance, microscale thermophoresis or dual polarisation interferometry to quantify the reaction affinities and kinetic properties and also any induced conformational change.

Chapter 6

Biological Membranes

A biological membrane or biomembrane is an enclosing or separating membrane that acts as a selectively permeable barrier within living things. Biological membranes, in the form of cell membranes, often consist of a phospholipid bilayer with embedded, integral and peripheral proteins used in communication and transportation of chemicals and ions. Bulk lipid in membrane provides a fluid matrix for proteins to rotate and laterally diffuse for physiological functioning. Proteins are adapted to high membrane fluidity environment of lipid bilayer with the presence of an annular lipid shell, consisting of lipid molecules bound tightly to surface of integral membrane proteins. The cellular membranes should not be confused with isolating tissues formed by layers of cells, such as mucous membranes and basement membranes.

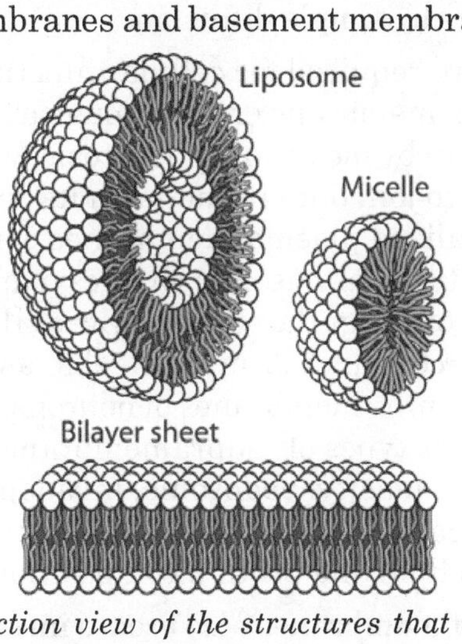

Figure: *Cross-section view of the structures that can be formed by phospholipids in aqueous*

Function

The phospholipid bilayer contains a charged hydrophilic head, which interacts with polar water. It also contains a hydrophobic tail, which meets with the hydrophobic tail of the complementary layer. This maintains the fluidity of the cell.

Membranes in cells typically define enclosed spaces or compartments in which cells may maintain a chemical or biochemical environment that differs from the outside. For example, the membrane around peroxisomes shields the rest of the cell from peroxides, chemicals that can be toxic to the cell, and the cell membrane separates a cell from its surrounding medium. Peroxisomes are one form of vacuole found in the cell that contain by-products of chemical reactions within the cell. Most organelles are defined by such membranes, and are called “membrane-bound” organelles.

Selective Permeability

Probably the most important feature of a biomembrane is that it is a selectively permeable structure. This means that the size, charge, and other chemical properties of the atoms and molecules attempting to cross it will determine whether they succeed in doing so. Selective permeability is essential for effective separation of a cell or organelle from its surroundings. Biological membranes also have certain mechanical or elastic properties that allow them to change shape and move as required.

Generally, small hydrophobic molecules can readily cross phospholipid bilayers by simple diffusion.

Particles that are required for cellular function but are unable to diffuse freely across a membrane enter through a membrane transport protein or are taken in by means of endocytosis, where the membrane allows for a vacuole to join onto it and push its contents into the cell. Many types of specialized plasma membranes can separate cell from external environment: apical, basolateral, presynaptic and postsynaptic ones, membranes of flagella, cilia, microvillus, filopodia and lamellipodia, the sarcolemma of muscle cells, as well as specialized myelin and dendritic spine membranes of neurons. Plasma membranes can also form different types of “supramembrane” structures such as caveola, postsynaptic density, podosome, invadopodium, desmosome, hemidesmosome, focal adhesion, and cell junctions. These types of membranes differ in lipid and protein composition.

Distinct types of membranes also create intracellular organelles: endosome; smooth and rough endoplasmic reticulum; sarcoplasmic reticulum; Golgi apparatus; lysosome; mitochondrion (inner and outer membranes); nucleus (inner and outer membranes); peroxisome; vacuole; cytoplasmic granules; cell vesicles (phagosome, autophagosome, clathrin-coated vesicles, COPI-coated and COPII-coated vesicles) and secretory vesicles (including synaptosome, acrosomes, melanosomes, and chromaffin granules). Different types of biological membranes have diverse lipid and protein compositions. The content of membranes defines their physical and biological properties. Some components of membranes play a key role in medicine, such as the efflux pumps that pump drugs out of a cell.

Nerve Signals

The nerve signal, or action potential, is a coordinated movement of sodium and potassium ions across the nerve cell membrane. Here's how it works:

As we discussed, the inside of the cell is slightly negatively charged (resting membrane potential of -70 to -80 mV).

A disturbance (mechanical, electrical, or sometimes chemical) causes a few sodium channels in a small portion of the membrane to open.

Sodium ions enter the cell through the open sodium channels. The positive charge that they carry makes the inside of the cell slightly less negative (depolarizes the cell).

When the depolarization reaches a certain threshold value, many more sodium channels in that area open. More sodium flows in and triggers an action potential. The inflow of sodium ions reverses the membrane potential in that area (making it positive inside and negative outside — the electrical potential goes to about +40 mV inside)

When the electrical potential reaches +40 mV inside (about 1 millisecond later), the sodium channels shut down and let no more sodium ions inside (sodium inactivation).

The developing positive membrane potential causes potassium channels to open.

Potassium ions leave the cell through the open potassium channels. The outward movement of positive potassium ions makes the inside of the membrane more negative and returns the membrane toward the resting membrane potential (repolarizes the cell).

When the membrane potential returns to the resting value, the potassium channels shut down and potassium ions can no longer leave the cell.

The membrane potential slightly overshoots the resting potential, which is corrected by the sodium-potassium pump, which restores the normal ion balance across the membrane and returns the membrane potential to its resting level.

Now, this sequence of events occurs in a local area of the membrane. But these changes get passed on to the next area of membrane, then to the next area, and so on down the entire length of the axon. Thus, the action potential (nerve impulse or nerve signal) gets transmitted (propagated) down the nerve cell.-

There are a few things to note about the propagation of the action potential. When an area has been depolarized and repolarized and the action potential has moved on to the next area, there is a short period of time before that first area can be depolarized again (refractory period). This refractory period prevents the action potential from moving backward and keeps everything moving in one direction.

- The action potential is an "all-or-none" response. Once the membrane reaches a threshold, it will depolarize to +40 mV. In other words, once the ionic events are set in motion, they will continue until the end.
- These ionic events occur in many excitable cells besides neurons (like muscle cells).
- Action potentials are propagated rapidly. Typical neurons conduct at 10 to 100 metres per second. Conduction speed varies with the diameter of the axon (larger = faster) and the presence of myelin (myelinated = faster). The rapid nerve conductions throughout the neural circuitry enable you to respond to stimuli in fractions of a second.
- The channels can be poisoned and prevented from opening. Various toxins (puffer fish toxin, snake venom, scorpion venom) can prevent specific channels from opening and distort the action potential or prevent it from happening altogether. Similarly, many local anesthetics (e.g. lidocaine, novocaine, benzocaine) can prevent action potentials from being propagated in the nerve cells in an area and temporarily prevent you from feeling pain.
- The propagation of the action potential is also sensitive to temperature in experimental settings. Colder temperatures slow down the action potential, but this usually doesn't happen in an individual. However, you can use cold-block techniques to temporarily anesthetize an area (like putting ice on an injured finger).

So, if the size of the action potential does not vary, how does an action potential code information? Information is encoded by the frequency of action potentials, much like FM radio. A small stimulus will initiate a low frequency train of a few action potentials. As the intensity of the stimulus increases, so does the frequency of action potentials.

Action Potential

Action potentials occur in several types of animal cells, called excitable cells, which include neurons, muscle cells, and endocrine cells, as well as in some plant cells. In neurons, they play a central role in cell-to-cell communication. In other types of cells, their main function is to activate intracellular processes. In muscle cells, for example, an action potential is the first step in the chain of events leading to contraction. In beta cells of the pancreas, they provoke release of insulin. Action potentials in neurons are also known as "nerve impulses" or "spikes", and the temporal sequence of action potentials generated by a neuron is called its "spike train". A neuron that emits an action potential is often said to "fire".

Action potentials are generated by special types of voltage-gated ion channels embedded in a cell's plasma membrane. These channels are shut when the membrane potential is near the resting potential of the cell, but they rapidly begin to open if the membrane potential increases to a precisely defined threshold value. When the channels open (in response to depolarization in transmembrane voltage), they allow an inward flow of sodium ions, which changes the electrochemical gradient, which in turn produces a further rise in the membrane potential. This then causes more channels to open, producing a greater electric current across the cell membrane, and so on. The process proceeds explosively until all of the available ion channels are open, resulting in a large upswing in the membrane potential. The rapid influx of sodium ions causes the polarity of the plasma membrane to reverse, and the ion channels then rapidly inactivate. As the sodium channels close, sodium ions can no longer enter the neuron, and then they are actively transported back out of the plasma membrane. Potassium channels are then activated, and there is an outward current of potassium ions, returning the electrochemical gradient to the resting state. After an action potential has occurred, there is a transient negative shift, called the afterhyperpolarization or refractory period, due to additional potassium currents. This is the mechanism that prevents an action potential from travelling back the way it just came.

In animal cells, there are two primary types of action potentials. One type is generated by voltage-gated sodium channels, the other by voltage-gated calcium channels. Sodium-based action potentials usually last for under one millisecond, whereas calcium-based action potentials may last for 100 milliseconds or longer. In some types of neurons, slow calcium spikes provide the driving force for a long burst of rapidly emitted sodium spikes. In cardiac muscle cells, on the other hand, an initial fast sodium spike provides a "primer" to provoke the rapid onset of a calcium spike, which then produces muscle contraction.

Function

Nearly all cell membranes in animals, plants and fungi maintain an electric potential difference (voltage)—the membrane potential. A typical voltage across an animal cell membrane is –65 mV—approximately one-fifteenth of a volt. Because the cell membrane is very thin, voltages of this magnitude give rise to very strong electric forces across the cell membrane.

In the majority of cells, the voltage stays fairly constant over time. There are some types of cells, however, that are electrically active in the sense that their voltages fluctuate. In some of these, the voltages sometimes show very rapid up-and-down fluctuations that have a stereotyped form: these up-and-down cycles are known as action potentials. The durations of action potentials vary across a wide range, and consequently they are analog signals. In brain cells of animals, the entire up-and-down cycle may take place in roughly a few thousandths of a second. In other types of cells, the cycle may last for several seconds.

The electrical properties of an animal cell are determined by the structure of the membrane that surrounds it. A cell membrane consists of a layer of lipid molecules with larger protein molecules embedded in it. The lipid layer is highly resistant to movement of electrically charged ions, so it functions mainly as an insulator. The large membrane-embedded molecules, in contrast, provide channels through which ions can pass across the membrane, and some of the large molecules are capable of actively moving specific types of ions from one side of the membrane to the other.

Process in a Typical Neuron

All cells in animal body tissues are electrically polarized – in other words, they maintain a voltage difference across the cell's plasma membrane, known as the membrane potential. This electrical polarization results from a complex interplay between protein structures embedded in the membrane called ion pumps and ion channels. In neurons, the types of ion channels in the membrane usually vary across different parts of the cell, giving the dendrites, axon, and cell body different electrical properties. As a result, some parts of the membrane of a neuron may be excitable (capable of generating action potentials), whereas others are not. Recent studies have shown that the most excitable part of a neuron is the part after the axon hillock (the point where the axon leaves the cell body), which is called the initial segment, but the axon and cell body are also excitable in most cases.

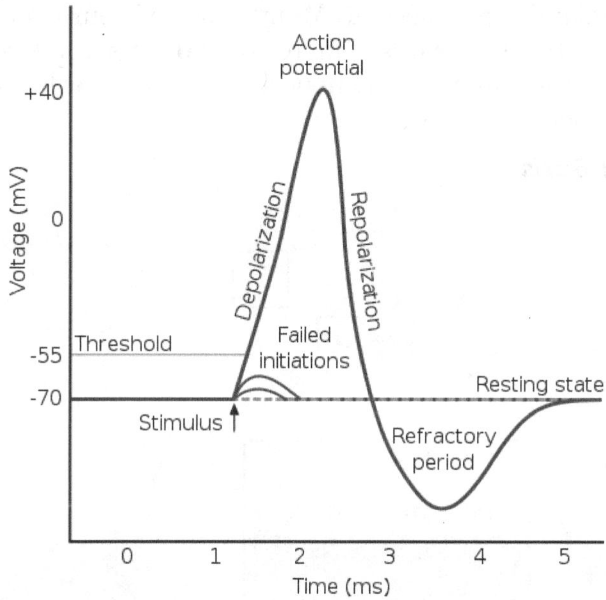

Figure**:** *Approximate plot of a typical action potential shows its various phases as the action potential passes a point on a cell membrane. The membrane potential starts out at -70 mV at time zero. A stimulus is applied at time = 1 ms, which raises the membrane potential above -55 mV (the threshold potential). After the stimulus is applied, the membrane potential rapidly rises to a peak potential of +40 mV at time = 2 ms. Just as quickly, the potential then drops and overshoots to -90 mV at time = 3 ms, and finally the resting potential of -70 mV is reestablished at time = 5 ms.*

Each excitable patch of membrane has two important levels of membrane potential: the resting potential, which is the value the membrane potential maintains as long as nothing perturbs the cell, and a higher value called the threshold potential. At the axon hillock of a typical neuron, the resting potential is around –70 millivolts (mV) and the threshold potential is around –55 mV. Synaptic inputs to a neuron cause the membrane to depolarize or hyperpolarize; that is, they cause the membrane potential to rise or fall. Action potentials are triggered when enough depolarization accumulates to bring the membrane potential up to threshold. When an action potential is triggered, the membrane potential abruptly shoots upward, often reaching as high as +100 mV, then equally abruptly shoots back downward, often ending below the resting level, where it remains for some period of time.

The shape of the action potential is stereotyped; that is, the rise and fall usually have approximately the same amplitude and time course for all action potentials in a given cell. (Exceptions are discussed later in the article.) In most neurons, the entire process takes place in about a thousandth of a second. Many types of neurons emit action potentials constantly at rates of up to 10–100 per second; some types, however, are much quieter, and may go for minutes or longer without emitting any action potentials.

Biophysical Basis

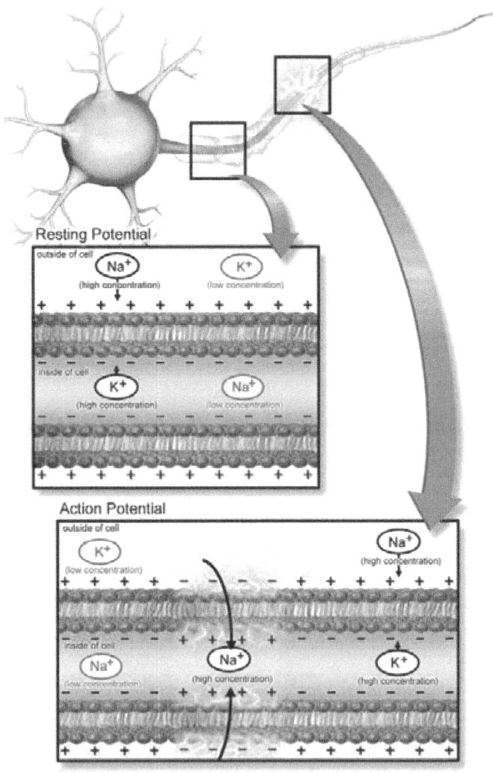

Figure: *Action potential propagation along an axon*

Action potentials result from the presence in a cell's membrane of special types of voltage-gated ion channels. A voltage-gated ion channel is a cluster of proteins embedded in the membrane that has three key properties:

1. It is capable of assuming more than one conformation.
2. At least one of the conformations creates a channel through the membrane that is permeable to specific types of ions.
3. The transition between conformations is influenced by the membrane potential.

Thus, a voltage-gated ion channel tends to be open for some values of the membrane potential, and closed for others. In most cases, however, the relationship between membrane potential and channel state is probabilistic and involves a time delay. Ion channels switch between conformations at unpredictable times: The membrane potential determines the rate of transitions and the probability per unit time of each type of transition.

Voltage-gated ion channels are capable of producing action potentials because they can give rise to positive feedback loops: The membrane potential controls the state of the ion channels, but the state of the ion channels controls the membrane potential. Thus, in some situations, a rise in the membrane potential can cause ion channels to open, thereby causing a further rise in the membrane potential. An action potential occurs when this positive feedback cycle proceeds explosively. The time and amplitude trajectory of the action potential are determined by the biophysical properties of the voltage-gated ion channels that produce it. Several types of channels that are capable of producing the positive feedback necessary to generate an action potential exist. Voltage-gated sodium channels are responsible for the fast action potentials involved in nerve conduction. Slower action potentials in muscle cells and some types of neurons are generated by voltage-gated calcium channels. Each of these types comes in multiple variants, with different voltage sensitivity and different temporal dynamics.

The most intensively studied type of voltage-dependent ion channels comprises the sodium channels involved in fast nerve conduction. These are sometimes known as Hodgkin-Huxley sodium channels because they were first characterized by Alan Hodgkin and Andrew Huxley in their Nobel Prize-winning studies of the biophysics of the action potential, but can more conveniently be referred to as Na_V channels. (The "V" stands for "voltage".) An Na_V channel has three possible states, known as *deactivated*, *activated*, and *inactivated*. The channel is permeable only to sodium ions when it is in the *activated* state. When the membrane potential is low, the channel spends most of its time in the *deactivated* (closed) state. If the membrane potential is raised above a certain level, the channel shows increased probability of transitioning to the *activated* (open) state. The higher the membrane potential the greater the probability of activation. Once a channel has activated, it will eventually transition to the *inactivated* (closed) state. It tends then to stay inactivated for some time, but, if the membrane potential becomes low again, the channel will eventually transition back to the *deactivated* state. During an action potential, most channels of this type go through a cycle *deactivated→activated→inactivated→ deactivated.* This is only the population average behaviour, however — an individual channel can in principle make any transition at any time. However, the likelihood of a channel's transitioning from the *inactivated* state directly to the *activated* state is very low: A channel in the *inactivated* state is refractory until it has transitioned back to the *deactivated* state.

The outcome of all this is that the kinetics of the Na_V channels are governed by a transition matrix whose rates are voltage-dependent in a complicated way. Since these channels themselves play a major role in determining the voltage, the global dynamics of the system can be quite difficult to work out. Hodgkin and Huxley approached the problem by developing a set of differential equations for the parameters that govern the ion channel states, known as the Hodgkin-Huxley equations. These equations have been extensively modified by later research, but form the starting point for most theoretical studies of action potential biophysics.

As the membrane potential is increased, sodium ion channels open, allowing the entry of sodium ions into the cell. This is followed by the opening of potassium ion channels that permit the exit of potassium ions from the cell. The inward flow of sodium ions increases the concentration of positively charged cations in the cell and causes depolarization, where the potential of the cell is higher than the cell's resting potential. The sodium channels close at the peak of the action potential, while potassium continues to leave the cell. The efflux of potassium ions decreases the membrane potential or hyperpolarizes the cell. For small voltage increases from rest, the potassium current exceeds the sodium current and the voltage returns to its normal resting value, typically "70 mV. However, if the voltage increases past a critical threshold, typically 15 mV higher than the resting value, the sodium current dominates. This results in a runaway condition whereby the positive feedback from the sodium current activates even more sodium channels. Thus, the cell *fires*, producing an action potential. The frequency at which cellular action potentials are produced is known as its *firing rate.*

Currents produced by the opening of voltage-gated channels in the course of an action potential are typically significantly larger than the initial stimulating current. Thus, the amplitude, duration, and shape of the action potential are determined largely by the properties of the excitable membrane and not the amplitude or duration of the stimulus. This all-or-nothing property of the action potential sets it apart from graded potentials such as receptor potentials, electrotonic potentials, and synaptic potentials, which scale with the magnitude of the stimulus.

A variety of action potential types exist in many cell types and cell compartments as determined by the types of voltage-gated channels, leak channels, channel distributions, ionic concentrations, membrane capacitance, temperature, and other factors.

The principal ions involved in an action potential are sodium and potassium cations; sodium ions enter the cell, and potassium ions leave, restoring equilibrium. Relatively few ions need to cross the membrane for the membrane voltage to change drastically. The ions exchanged during an action potential, therefore, make a negligible change in the interior and exterior ionic concentrations. The few ions that do cross are pumped out again by the continuous action of the sodium–potassium pump, which, with other ion transporters, maintains the normal ratio of ion concentrations across the membrane. Calcium cations and chloride anions are involved in a few types of action potentials, such as the cardiac action potential and the action potential in the single-cell alga *Acetabularia*, respectively.

Although action potentials are generated locally on patches of excitable membrane, the resulting currents can trigger action potentials on neighbouring stretches of membrane, precipitating a domino-like propagation. In contrast to passive spread of electric potentials (electrotonic potential), action potentials are generated anew along excitable stretches of membrane and propagate without decay. Myelinated sections of axons are not excitable and do not produce action potentials and the signal is propagated passively as electrotonic potential. Regularly spaced unmyelinated patches, called the nodes of Ranvier, generate action potentials to boost the signal.

Known as saltatory conduction, this type of signal propagation provides a favourable tradeoff of signal velocity and axon diameter. Depolarization of axon terminals, in general, triggers the release of neurotransmitter into the synaptic cleft. In addition, backpropagating action potentials have been recorded in the dendrites of pyramidal neurons, which are ubiquitous in the neocortex. These are thought to have a role in spike-timing-dependent plasticity.

Neurotransmission

Several types of cells support an action potential, such as plant cells, muscle cells, and the specialized cells of the heart (in which occurs the cardiac action potential). However, the main excitable cell is the neuron, which also has the simplest mechanism for the action potential.

Neurons are electrically excitable cells composed, in general, of one or more dendrites, a single soma, a single axon and one or more axon terminals. Dendrites are cellular projections whose primary function is to receive synaptic signals. Their protrusions, or spines, are designed to capture the neurotransmitters released by the presynaptic neuron. They have a high concentration of ligand-gated ion channels. These spines have a thin neck connecting a bulbous protrusion to the dendrite. This ensures that changes occurring inside the spine are less likely to affect the neighbouring spines. The dendritic spine can, with rare exception, act as an independent unit. The dendrites extend from the soma, which houses the nucleus, and many of the "normal" eukaryotic organelles.

Unlike the spines, the surface of the soma is populated by voltage activated ion channels. These channels help transmit the signals generated by the dendrites. Emerging out from the soma is the axon hillock. This region is characterized by having a very high concentration of voltage-activated sodium channels. In general, it is considered to be the spike initiation zone for action potentials. Multiple signals generated at the spines, and transmitted by the soma all converge here. Immediately after the axon hillock is the axon. This is a thin tubular protrusion travelling away from the soma. The axon is insulated by a myelin sheath. Myelin is composed of either Schwann cells (in the peripheral nervous system) or oligodendrocytes (in the central nervous system), both of which are types of glial cells.

Although glial cells are not involved with the transmission of electrical signals, they communicate and provide important biochemical support to neurons. To be specific, myelin wraps multiple times around the axonal segment, forming a thick fatty layer that prevents ions from entering or escaping the axon. This insulation prevents significant signal decay as well as ensuring faster signal speed. This insulation, however, has the restriction that no channels can be present on the surface of the axon. There are, therefore, regularly spaced patches of membrane, which have no insulation. These nodes of Ranvier can be considered to be "mini axon hillocks", as their purpose is to boost the signal in order to prevent significant signal decay. At the furthest end, the axon loses its insulation and begins to branch into several axon terminals. These presynaptic terminals, or synaptic boutons, are a specialized area within the axon of the presynaptic cell that contains neurotransmitters enclosed in small membrane-bound spheres called synaptic vesicles.

Initiation

Before considering the propagation of action potentials along axons and their termination at the synaptic knobs, it is helpful to consider the methods by which action potentials can be initiated at the axon hillock. The basic requirement is that the membrane voltage at the hillock be raised above the threshold for firing. There are several ways in which this depolarization can occur.

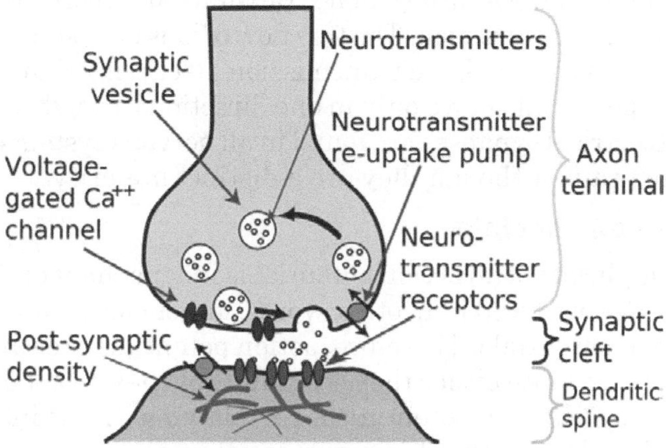

Figure: *When an action potential arrives at the end of the pre-synaptic axon (yellow), it causes the release of neurotransmitter molecules that open ion channels in the post-synaptic neuron (green). The combined excitatory and inhibitory postsynaptic potentials of such inputs can begin a new action potential in the post-synaptic neuron.*

Dynamics

Action potentials are most commonly initiated by excitatory postsynaptic potentials from a presynaptic neuron. Typically, neurotransmitter molecules are released by the presynaptic neuron. These neurotransmitters then bind to receptors on the postsynaptic cell. This binding opens various types of ion channels. This opening has the further effect of changing the local permeability of the cell membrane and, thus, the membrane potential. If the binding increases the voltage (depolarizes the membrane), the synapse is excitatory. If, however, the binding decreases the voltage (hyperpolarizes the membrane), it is inhibitory. Whether the voltage is increased or decreased, the change propagates passively to nearby regions of the membrane (as described by the cable equation and its refinements). Typically, the voltage stimulus decays exponentially with the distance from the synapse and with time from the binding of the neurotransmitter. Some fraction of an excitatory voltage may reach the axon hillock and may (in rare cases) depolarize the membrane enough to provoke a new action potential. More typically, the excitatory potentials from several synapses must work together at nearly the same time to provoke a new action potential. Their joint efforts can be thwarted, however, by the counteracting inhibitory postsynaptic potentials.

Neurotransmission can also occur through electrical synapses. Due to the direct connection between excitable cells in the form of gap junctions, an action potential can be transmitted directly from one cell to the next in either direction. The free flow of ions between cells enables rapid non-chemical-mediated transmission. Rectifying channels ensure that action potentials move only in one direction through an electrical synapse. Electrical synapses are found in all nervous systems, including the human brain, although they are a distinct minority.

"All-or-None" Principle

The amplitude of an action potential is independent of the amount of current that produced it. In other words, larger currents do not create larger action potentials. Therefore, action potentials are said to be all-or-none signals, since either they occur fully or they do not occur at all. The frequency of action potentials is correlated with the intensity of a stimulus. This is in contrast to receptor potentials, whose amplitudes are dependent on the intensity of a stimulus.

Sensory Neurons

In sensory neurons, an external signal such as pressure, temperature, light, or sound is coupled with the opening and closing of ion channels, which in turn alter the ionic permeabilities of the membrane and its voltage. These voltage changes can again be excitatory (depolarizing) or inhibitory (hyperpolarizing) and, in some sensory neurons, their combined effects can depolarize the axon hillock enough to provoke action potentials. Examples in humans include the olfactory receptor neuron and Meissner's corpuscle, which are critical for the sense of smell and touch, respectively. However, not all sensory neurons convert their external signals into action potentials; some do not even have an axon! Instead, they may convert the signal into the release of a neurotransmitter, or into continuous graded potentials, either of which may stimulate subsequent neuron(s) into firing an action potential. For illustration, in the human ear, hair cells convert the incoming sound into the opening and closing of mechanically gated ion channels, which may cause neurotransmitter molecules to be released. In similar manner, in the human retina, the initial photoreceptor cells and the next layer of cells (comprising bipolar cells and horizontal cells) do not produce action potentials; only some amacrine cells and the third layer, the ganglion cells, produce action potentials, which then travel up the optic nerve.

Pacemaker Potentials

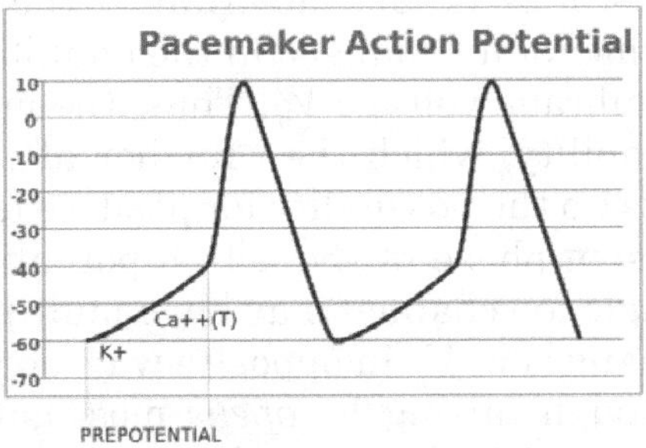

Figure: *In pacemaker potentials, the cell spontaneously depolarizes (straight line with upward slope) until it fires an action potential.*

In sensory neurons, action potentials result from an external stimulus. However, some excitable cells require no such stimulus to fire: They spontaneously depolarize their axon hillock and fire action potentials at a regular rate, like an internal clock. The voltage traces of such cells are known as pacemaker potentials. The cardiac pacemaker cells of the sinoatrial node in the heart provide a good example. Although such pacemaker potentials have a natural rhythm, it can be adjusted by external stimuli; for instance, heart rate can be altered by pharmaceuticals as well as signals from the sympathetic and parasympathetic nerves. The external stimuli do not cause the cell's repetitive firing, but merely alter its timing. In some cases, the regulation of frequency can be more complex, leading to patterns of action potentials, such as bursting.

Phases

The course of the action potential can be divided into five parts: the rising phase, the peak phase, the falling phase, the undershoot phase, and the refractory period. During the rising phase the membrane potential depolarizes (becomes more positive). The point at which depolarization stops is called the peak phase. At this stage, the membrane potential reaches a maximum. Subsequent to this, there is a falling phase. During this stage the membrane potential becomes more negative, returning towards resting potential. The undershoot, or afterhyperpolarization, phase is the period during which the membrane potential temporarily becomes more negatively charged than when at rest (hyperpolarized). Finally, the time during which a subsequent action potential is impossible or difficult to fire is called the refractory period, which may overlap with the other phases.

The course of the action potential is determined by two coupled effects. First, voltage-sensitive ion channels open and close in response to changes in the membrane voltage V_m. This changes the membrane's permeability to those ions. Second, according to the Goldman equation, this change in permeability changes in the equilibrium potential E_m, and, thus, the membrane voltage V_m. Thus, the membrane potential affects the permeability, which then further affects the membrane potential. This sets up the possibility for positive feedback, which is a key part of the rising phase of the action potential. A complicating factor is that a single ion channel may have multiple internal "gates" that respond to changes in V_m in opposite ways, or at different rates. For example, although raising V_m *opens* most gates in the voltage-sensitive sodium channel, it also *closes* the channel's "inactivation gate", albeit more slowly. Hence, when V_m is raised suddenly, the sodium channels open initially, but then close due to the slower inactivation.

The voltages and currents of the action potential in all of its phases were modelled accurately by Alan Lloyd Hodgkin and Andrew Huxley in 1952, for which they were awarded the Nobel Prize in Physiology or Medicine in 1963. However, their model considers only two types of voltage-sensitive ion channels, and makes several assumptions about them, e.g., that their internal gates open and close independently of one another. In reality, there are many types of ion channels, and they do not always open and close independently.

Stimulation and Rising Phase

A typical action potential begins at the axon hillock with a sufficiently strong depolarization, e.g., a stimulus that increases V_m. This depolarization is often caused by the injection of extra sodium cations into the cell; these cations can come from a wide variety of sources, such as chemical synapses, sensory neurons or pacemaker potentials.

For a neuron at rest, there is a high concentration of sodium and chlorine ions in the extracellular fluid compared to the intracellular fluid while there is a high concentration of potassium ions in the intracellular fluid compared to the extracellular fluid. This concentration gradient along with potassium leak channels present on the membrane of the neuron causes an efflux of potassium ions making the resting potential close to E_KH" –75 mV. The depolarization opens both the sodium and potassium channels in the membrane, allowing the ions to flow into and out of the axon, respectively. If the depolarization is small (say, increasing V_m from –70 mV to –60 mV), the outward potassium current overwhelms the inward sodium current and the membrane repolarizes back to its normal resting potential around –70 mV.However, if the depolarization is large enough, the inward sodium current increases more than the outward potassium current and a runaway condition (positive feedback) results: the more inward current there is, the more V_m increases, which in turn further increases the inward current. A sufficiently strong depolarization (increase in V_m) causes the voltage-sensitive sodium channels to open; the increasing permeability to sodium drives V_m closer to the sodium equilibrium voltage E_{Na}H" +55 mV. The increasing voltage in turn causes even more sodium channels to open, which pushes V_m still further towards E_{Na}. This positive feedback continues until the sodium channels are fully open and V_m is close to E_{Na}. The sharp rise in V_m and sodium permeability correspond to the *rising phase* of the action potential.

The critical threshold voltage for this runaway condition is usually around –45 mV, but it depends on the recent activity of the axon. A membrane that has just fired an action potential cannot fire another one immediately, since the ion channels have not returned to the deactivated state. The period during which no new action potential can be fired is called the *absolute refractory period*. At longer times, after some but not all of the ion channels have recovered, the axon can be stimulated to produce another action potential, but with a higher threshold, requiring a much stronger depolarization, e.g., to –30 mV. The period during which action potentials are unusually difficult to evoke is called the *relative refractory period*.

Peak and Falling Phase

The positive feedback of the rising phase slows and comes to a halt as the sodium ion channels become maximally open. At the peak of the action potential, the sodium permeability is maximized and the membrane voltage V_m is nearly equal to the sodium equilibrium voltage E_{Na}. However, the same raised voltage that opened the sodium channels initially also slowly shuts them off, by closing their pores; the sodium channels become *inactivated*. This lowers the membrane's permeability to sodium relative to potassium, driving the membrane voltage back towards the resting value. At the same time, the raised voltage opens voltage-sensitive potassium channels; the increase in the membrane's potassium permeability drives V_m towards E_K. Combined, these changes in sodium and potassium permeability cause V_m to drop quickly, repolarizing the membrane and producing the "falling phase" of the action potential.

Afterhyperpolarization

The raised voltage opened many more potassium channels than usual, and some of these do not close right away when the membrane returns to its normal resting voltage. In addition, further potassium channels open in response to the influx of calcium ions during the action potential. The potassium permeability of the membrane is transiently unusually high, driving the membrane voltage V_m even closer to the potassium equilibrium voltage E_K. Hence, there is an undershoot or hyperpolarization, termed an afterhyperpolarization in technical language, that persists until the membrane potassium permeability returns to its usual value.

Refractory Period

Each action potential is followed by a refractory period, which can be divided into an *absolute refractory period*, during which it is impossible to evoke another action potential, and then a *relative refractory period*, during which a stronger-than-usual stimulus is required. These two refractory periods are caused by changes in the state of sodium and potassium channel molecules. When closing after an action potential, sodium channels enter an "inactivated" state, in which they cannot be made to open regardless of the membrane potential—this gives rise to the absolute refractory period. Even after a sufficient number of sodium channels have transitioned back to their resting state, it frequently happens that a fraction of potassium channels remains open, making it difficult for the membrane potential to depolarize, and thereby giving rise to the relative refractory period. Because the density and subtypes of potassium channels may differ greatly between different types of neurons, the duration of the relative refractory period is highly variable.

The absolute refractory period is largely responsible for the unidirectional propagation of action potentials along axons. At any given moment, the patch of axon behind the actively spiking part is refractory, but the patch in front, not having been activated recently, is capable of being stimulated by the depolarization from the action potential.

Propagation

The action potential generated at the axon hillock propagates as a wave along the axon. The currents flowing inwards at a point on the axon during an action potential spread out along the axon, and depolarize the adjacent sections of its membrane. If sufficiently strong, this depolarization provokes a similar action potential at the neighbouring membrane patches. This basic mechanism was demonstrated by Alan Lloyd Hodgkin in 1937. After crushing or cooling nerve segments and thus blocking the action potentials, he showed that an action potential arriving on one side of the block could provoke another action potential on the other, provided that the blocked segment was sufficiently short.

Once an action potential has occurred at a patch of membrane, the membrane patch needs time to recover before it can fire again. At the molecular level, this *absolute refractory period* corresponds to the time required for the voltage-activated sodium channels to recover from inactivation, i.e., to return to their closed state.

There are many types of voltage-activated potassium channels in neurons, some of them inactivate fast (A-type currents) and some of them inactivate slowly or not inactivate at all; this variability guarantees that there will be always an available source of current for repolarization, even if some of the potassium channels are inactivated because of preceding depolarization. On the other hand, all neuronal voltage-activated sodium channels inactivate within several millisecond during strong depolarization, thus making following depolarization impossible until a substantial fraction of sodium channels have returned to their closed state.

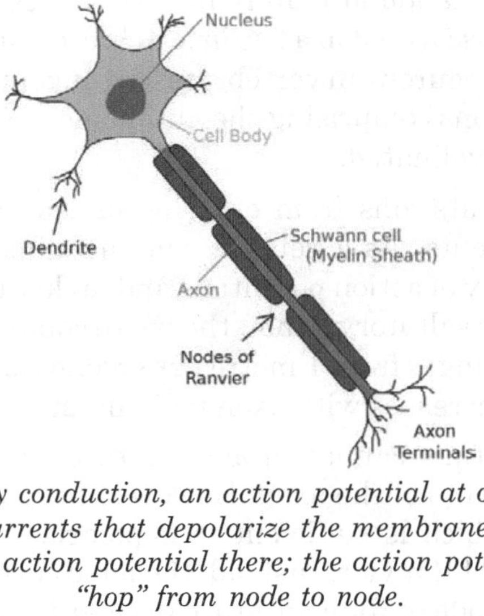

Figure: *In saltatory conduction, an action potential at one node of Ranvier causes inwards currents that depolarize the membrane at the next node, provoking a new action potential there; the action potential appears to "hop" from node to node.*

Although it limits the frequency of firing, the absolute refractory period ensures that the action potential moves in only one direction along an axon. The currents flowing in due to an action potential spread out in both directions along the axon. However, only the unfired part of the axon can respond with an action potential; the part that has just fired is unresponsive until the action potential is safely out of range and cannot restimulate that part.

In the usual orthodromic conduction, the action potential propagates from the axon hillock towards the synaptic knobs (the axonal termini); propagation in the opposite direction—known as antidromic conduction—is very rare. However, if a laboratory axon is stimulated in its middle, both halves of the axon are "fresh", i.e., unfired; then two action potentials will be generated, one travelling towards the axon hillock and the other travelling towards the synaptic knobs.

Myelin and Saltatory Conduction

In order to enable fast and efficient transduction of electrical signals in the nervous system, certain neuronal axons are covered with myelin sheaths. Myelin is a multilamellar membrane that enwraps the axon in segments separated by intervals known as nodes of Ranvier. It is produced by specialized cells: Schwann cells exclusively in the peripheral nervous system, and oligodendrocytes exclusively in the central nervous system. Myelin sheath reduces membrane capacitance and increases membrane resistance in the inter-node intervals, thus allowing a fast, saltatory movement of action potentials from node to node. Myelination is found mainly in vertebrates, but an analogous system has been discovered in a few invertebrates, such as some species of shrimp. Not all neurons in vertebrates are myelinated; for example, axons of the neurons comprising the autonomous nervous system are not, in general, myelinated.

Myelin prevents ions from entering or leaving the axon along myelinated segments. As a general rule, myelination increases the conduction velocity of action potentials and makes them more energy-efficient. Whether saltatory or not, the mean conduction velocity of an action potential ranges from 1 metre per second (m/s) to over 100 m/s, and, in general, increases with axonal diameter.

Action potentials cannot propagate through the membrane in myelinated segments of the axon. However, the current is carried by the cytoplasm, which is sufficient to depolarize the first or second subsequent node of Ranvier. Instead, the ionic current from an action potential at one node of Ranvier provokes another action potential at the next node; this apparent "hopping" of the action potential from node to node is known as saltatory conduction. Although the mechanism of saltatory conduction was suggested in 1925 by Ralph Lillie, the first experimental evidence for saltatory conduction came from Ichiji Tasaki and Taiji Takeuchi and from Andrew Huxley and Robert Stδmpfli. By contrast, in unmyelinated axons, the action potential provokes another in the membrane immediately adjacent, and moves continuously down the axon like a wave.

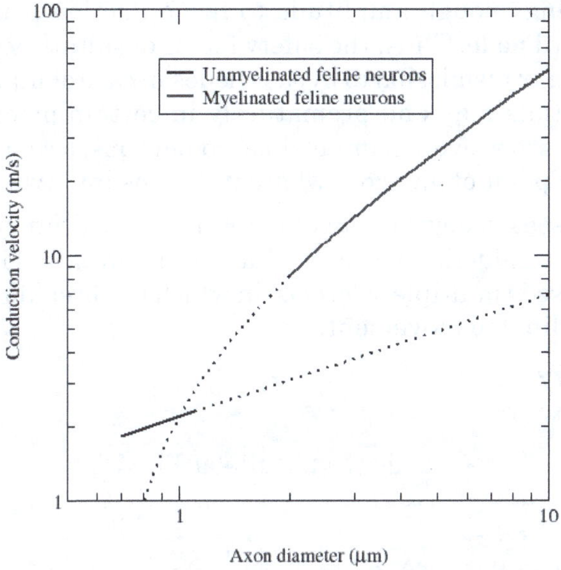

Figure: *Comparison of the conduction velocities of myelinated and unmyelinated axons in the cat. The conduction velocity v of myelinated neurons varies roughly linearly with axon diameter d (that is, v ∝ d), whereas the speed of unmyelinated neurons varies roughly as the square root (v ∝ √ d). The red and blue curves are fits of experimental data, whereas the dotted lines are their theoretical extrapolations.*

Myelin has two important advantages: fast conduction speed and energy efficiency. For axons larger than a minimum diameter (roughly 1 micrometre), myelination increases the conduction velocity of an action potential, typically tenfold. Conversely, for a given conduction velocity, myelinated fibers are smaller than their unmyelinated counterparts. For example, action potentials move at roughly the same speed (25 m/s) in a myelinated frog axon and an unmyelinated squid giant axon, but the frog axon has a roughly 30-fold smaller diameter and 1000-fold smaller cross-sectional area. Also, since the ionic currents are confined to the nodes of Ranvier, far fewer ions "leak" across the membrane, saving metabolic energy. This saving is a significant selective advantage, since the human nervous system uses approximately 20% of the body's metabolic energy.

The length of axons' myelinated segments is important to the success of saltatory conduction. They should be as long as possible to maximize the speed of conduction, but not so long that the arriving signal is too weak to provoke an action potential at the next node of Ranvier. In nature, myelinated segments are generally long enough for the passively propagated signal to travel for at least two nodes while retaining enough amplitude to fire an action potential at the second or third node. Thus, the safety factor of saltatory conduction is high, allowing transmission to bypass nodes in case of injury. However, action potentials may end prematurely in certain places where the safety factor is low, even in unmyelinated neurons; a common example is the branch point of an axon, where it divides into two axons.

Some diseases degrade myelin and impair saltatory conduction, reducing the conduction velocity of action potentials. The most well-known of these is multiple sclerosis, in which the breakdown of myelin impairs coordinated movement.

Cable Theory

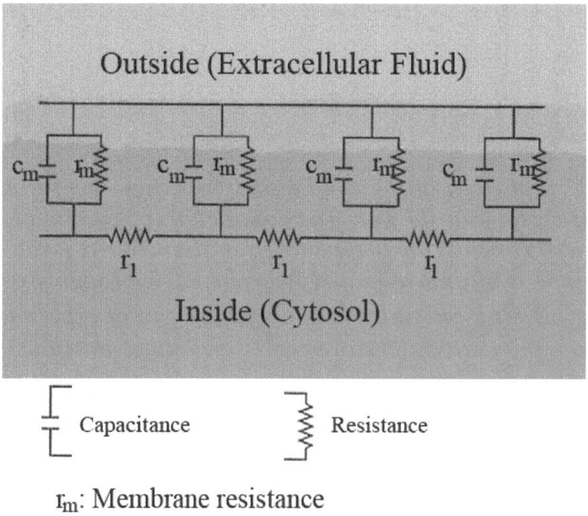

r_m: Membrane resistance
r_l : Longitudinal resistance
c_m: Capacitance due to electrostatic forces

***Figure**: Cable theory's simplified view of a neuronal fiber. The connected RC circuits correspond to adjacent segments of a passive neurite. The extracellular resistances r_e (the counterparts of the intracellular resistances r_i) are not shown, since they are usually negligibly small; the extracellular medium may be assumed to have the same voltage everywhere.*

The flow of currents within an axon can be described quantitatively by cable theory and its elaborations, such as the compartmental model.

Cable theory was developed in 1855 by Lord Kelvin to model the transatlantic telegraph cable and was shown to be relevant to neurons by Hodgkin and Rushton in 1946. In simple cable theory, the neuron is treated as an electrically passive, perfectly cylindrical transmission cable, which can be described by a partial differential equation

$$\tau \frac{\partial V}{\partial t} = \lambda^2 \frac{\partial^2 V}{\partial x^2} - V$$

where $V(x, t)$ is the voltage across the membrane at a time t and a position x along the length of the neuron, and where λ and τ are the characteristic length and time scales on which those voltages decay in response to a stimulus. Referring to the circuit diagram on the right, these scales can be determined from the resistances and capacitances per unit length.

$$\tau = r_m c_m$$

$$\lambda = \sqrt{\frac{r_m}{r_l}}$$

These time and length-scales can be used to understand the dependence of the conduction velocity on the diameter of the neuron in unmyelinated fibers. For example, the time-scale τ increases with both the membrane resistance r_m and capacitance c_m. As the capacitance increases, more charge must be transferred to produce a given transmembrane voltage (by the equation $Q=CV$); as the resistance increases, less charge is transferred per unit time, making the equilibration slower. In similar manner, if the internal resistance per unit length r_i is lower in one axon than in another (e.g., because the radius of the former is larger), the spatial decay length λ becomes longer and the conduction velocity of an action potential should increase. If the transmembrane resistance r_m is increased, that lowers the average "leakage" current across the membrane, likewise causing λ to become longer, increasing the conduction velocity.

Termination

Chemical Synapses

In general, action potentials that reach the synaptic knobs cause a neurotransmitter to be released into the synaptic cleft. Neurotransmitters are small molecules that may open ion channels in the postsynaptic cell; most axons have the same neurotransmitter at all of their termini. The arrival of the action potential opens voltage-sensitive calcium channels in the presynaptic membrane; the influx of calcium causes vesicles filled with neurotransmitter to migrate to the cell's surface and release their contents into the synaptic cleft. This complex process is inhibited by the neurotoxins tetanospasmin and botulinum toxin, which are responsible for tetanus and botulism, respectively.

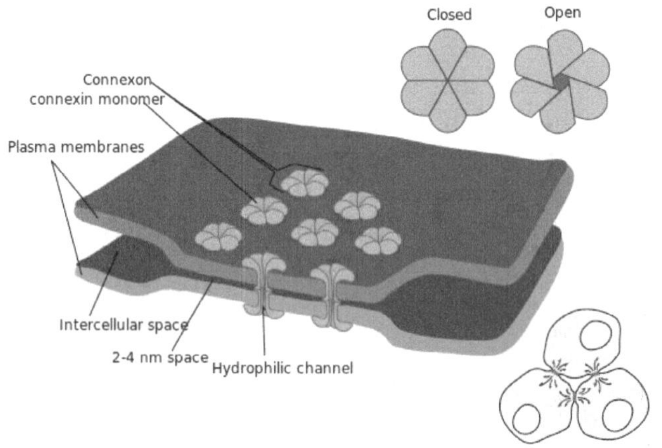

Electrical synapses between excitable cells allow ions to pass directly from one cell to another, and are much faster than chemical synapses.

Electrical Synapses

Some synapses dispense with the "middleman" of the neurotransmitter, and connect the presynaptic and postsynaptic cells together. When an action potential reaches such a synapse, the ionic currents flowing into the presynaptic cell can cross the barrier of the two cell membranes and enter the postsynaptic cell through pores known as connexons. Thus, the ionic currents of the presynaptic action potential can directly stimulate the postsynaptic cell. Electrical synapses allow for faster transmission because they do not require the slow diffusion of neurotransmitters across the synaptic cleft. Hence, electrical synapses are used whenever fast response and coordination of timing are crucial, as in escape reflexes, the retina of vertebrates, and the heart.

Neuromuscular Junctions

A special case of a chemical synapse is the neuromuscular junction, in which the axon of a motor neuron terminates on a muscle fiber. In such cases, the released neurotransmitter is acetylcholine, which binds to the acetylcholine receptor, an integral membrane protein in the membrane (the *sarcolemma*) of the muscle fiber. However, the acetylcholine does not remain bound; rather, it dissociates and is hydrolyzed by the enzyme, acetylcholinesterase, located in the synapse. This enzyme quickly reduces the stimulus to the muscle, which allows the degree and timing of muscular contraction to be regulated delicately. Some poisons inactivate acetylcholinesterase to prevent this control, such as the nerve agents sarin and tabun, and the insecticides diazinon and malathion.

Other Cell Types

Cardiac Action Potentials

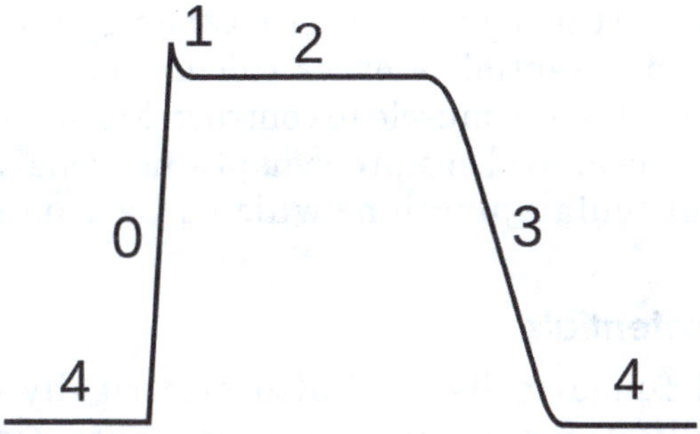

Figure: *Phases of a cardiac action potential. The sharp rise in voltage ("0") corresponds to the influx of sodium ions, whereas the two decays ("1" and "3", respectively) correspond to the sodium-channel inactivation and the repolarizing eflux of potassium ions. The characteristic plateau ("2") results from the opening of voltage-sensitive calcium channels.*

The cardiac action potential differs from the neuronal action potential by having an extended plateau, in which the membrane is held at a high voltage for a few hundred milliseconds prior to being repolarized by the potassium current as usual. This plateau is due to the action of slower calcium channels opening and holding the membrane voltage near their equilibrium potential even after the sodium channels have inactivated.

The cardiac action potential plays an important role in coordinating the contraction of the heart. The cardiac cells of the sinoatrial node provide the pacemaker potential that synchronizes the heart. The action potentials of those cells propagate to and through the atrioventricular node (AV node), which is normally the only conduction pathway between the atria and the ventricles. Action potentials from the AV node travel through the bundle of His and thence to the Purkinje fibers. Conversely, anomalies in the cardiac action potential—whether due to a congenital mutation or injury—can lead to human pathologies, especially arrhythmias. Several anti-arrhythmia drugs act on the cardiac action potential, such as quinidine, lidocaine, beta blockers, and verapamil.

Muscular Action Potentials

The action potential in a normal skeletal muscle cell is similar to the action potential in neurons. Action potentials result from the depolarization of the cell membrane (the sarcolemma), which opens voltage-sensitive sodium channels; these become inactivated and the membrane is repolarized through the outward current of potassium ions. The resting potential prior to the action potential is typically "90mV, somewhat more negative than typical neurons. The muscle action potential lasts roughly 2–4 ms, the absolute refractory period is roughly 1–3 ms, and the conduction velocity along the muscle is roughly 5 m/s. The action potential releases calcium ions that free up the tropomyosin and allow the muscle to contract. Muscle action potentials are provoked by the arrival of a pre-synaptic neuronal action potential at the neuromuscular junction, which is a common target for neurotoxins.

Plant Action Potentials

Plant and fungal cells are also electrically excitable. The fundamental difference to animal action potentials is that the depolarization in plant cells is not accomplished by an uptake of positive sodium ions, but by release of negative *chloride* ions. Together with the following release of positive potassium ions, which is common to plant and animal action potentials, the action potential in plants infers, therefore, an osmotic loss of salt (KCl), whereas the animal action potential is osmotically neutral, when equal amounts of entering sodium and leaving potassium cancel each other osmotically. The interaction of electrical and osmotic relations in plant cells indicates an osmotic function of electrical excitability in the common, unicellular ancestors of plants and animals under changing salinity conditions, whereas the present function of rapid signal transmission is seen as a younger accomplishment of metazoan cells in a more stable osmotic environment. It must be assumed that the familiar signalling function of action potentials in some vascular plants (e.g. *Mimosa pudica*), arose independently from that in metazoan excitable cells.

Taxonomic Distribution and Evolutionary Advantages

Action potentials are found throughout multicellular organisms, including plants, invertebrates such as insects, and vertebrates such as reptiles and mammals. Sponges seem to be the main phylum of multicellular eukaryotes that does not transmit action potentials, although some studies have suggested that these organisms have a form of electrical signaling, too. The resting potential, as well as the size and duration of the action potential, have not varied much with evolution, although the conduction velocity does vary dramatically with axonal diameter and myelination.

Comparison of action potentials (APs) from a representative cross-section of animals[48]					
Animal	***Cell type***	***Resting potential (mV)***	***AP increase (mV)***	***AP duration (ms)***	***Conduction speed (m/s)***
Squid (*Loligo*)	Giant axon	–60	120	0.75	35
Earthworm (*Lumbricus*)	Median giant fiber	–70	100	1.0	30
Cockroach (*Periplaneta*)	Giant fiber	–70	80–104	0.4	10
Frog (*Rana*)	Sciatic nerve axon	–60 to –80	110–130	1.0	7–30
Cat (*Felis*)	Spinal motor neuron	–55 to –80	80–110	1–1.5	30–120

Given its conservation throughout evolution, the action potential seems to confer evolutionary advantages. One function of action potentials is rapid, long-range signaling within the organism; the conduction velocity can exceed 110 m/s, which is one-third the speed of sound. For comparison, a hormone molecule carried in the bloodstream moves at roughly 8 m/s in large arteries. Part of this function is the tight coordination of mechanical events, such as the contraction of the heart. A second function is the computation associated with its generation. Being an all-or-none signal that does not decay with transmission distance, the action potential has similar advantages to digital electronics. The integration of various dendritic signals at the axon hillock and its thresholding to form a complex train of action potentials is another form of computation, one that has been exploited biologically to form central pattern generators and mimicked in artificial neural networks.

Memory

In biology, memory is present if the state of a biological system depends on its history in addition to present conditions. If this memory is recorded in the genetic material and stably inherited through cell division (mitosis or meiosis), it is genetic memory.

Somatic Memory

Somatic memory is limited to the organism and not passed on to subsequent generations. However, its mechanism may involve mitotically stable genetic memory. The term applies for cellular memory, animals' memory, and plants' memory, as described in the following paragraphs.

Cellular Memory

All cells in multicellular organisms are derived from a pluripotent zygote and contain the same genetic material (with a few exceptions). However, they are capable of recording a history of their development within the organism leading to their specialized functions and limitations. Cells often employ epigenetic processes that affect DNA-protein interactions to record this *cellular memory* in the form of mitotically stable changes of the genetic material without a change in the DNA sequence itself. This is typically achieved via changes of the chromatin structure. Examples are methylation patterns of the DNA molecule itself and proteins involved in packaging DNA, such as histones (also referred to as "histone code").

In Animals

A case of somatic genetic memory is the *immunological memory* of the adaptive immune response in vertebrates. The immune system is capable of learning to recognize pathogens and keeping a memory of this learning process, which is the basis of the success of vaccinations. Antibody genes in B and T lymphocytes are assembled from separate gene segments, giving each lymphocyte a unique antibody coding sequence leading to the vast diversity of antibodies in the immune system. If stimulated by an antigen (e.g. following vaccination or an infection with a pathogen), these antibodies are further fine-tuned via hypermutation. Memory B cells capable of producing these antibodies form the basis for acquired immunological memory. Each individual therefore carries a unique genetic memory of its immune system's close encounters with pathogens. As a somatic memory, this is not passed on to the next generation.

In Plants

Plants that undergo vernalization (promotion of flowering by a prolonged exposure to cold temperatures) record a *genetic memory of winter* to gain the competence to flower. The process involves epigenetically recording the length of cold exposure through chromatin remodelling which leads to mitotically stable changes in gene expression (the "winter code"). This releases the inhibition of flowering initiation and allows the plants to bloom with the correct timing at the onset of spring. As a somatic memory, this state is not passed on to subsequent generations but has to be acquired by each individual plant. The process of vernalization was falsely assumed to be a stably inherited genetic memory passed on to subsequent generations by the Russian geneticist Trofim Lysenko. Lysenko's claims of genetic memory and efforts to obtain or fabricate results in proof of it had disastrous effects for Russian genetics in the early 20th century (also see: Lysenkoism).

Inherited Epigenetic Memory

In genetics, genomic imprinting or other patterns of inheritance that are not determined by DNA sequence alone can form an epigenetic memory that is passed on to subsequent generations through meiosis. In contrast, somatic genetic memories are passed on by mitosis and limited to the individual, but are not passed on to the offspring. Both processes include similar epigenetic mechanisms, e.g. involving histones and methylation patterns.

Microbial Memory

In microbes, genetic memory is present in the form of inversion of specific DNA sequences serving as a switch between alternative patterns of gene expression.

Evolution

In population genetics and evolution, genetic memory represents the recorded history of adaptive changes in a species. Selection of organisms carrying genes coding for the best adapted proteins results in the evolution of species. An example for such a genetic memory is the innate immune response that represents a recording of the history of common microbial and viral pathogens encountered throughout the evolutionary history of the species. In contrast to the somatic memory of the adaptive immune response, the innate immune response is present at birth and does not require the immune system to learn to recognize antigens.

In the history of theories of evolution, the proposed genetic memory of an individual's experiences and environmental influences was a central part of Lamarckism to explain the inheritance of evolutionary changes.

Animal Behaviour

In ethology, genetic memory refers the inheritance of instinct in animals and humans.

Control of Movement

Motor control is the process by which humans and animals use their neuromuscular system to activate and coordinate the muscles and limbs involved in the performance of a motor skill. Fundamentally, it is the integration of sensory information, both about the world and the current state of the body, to determine the appropriate set of muscle forces and joint activations to generate some desired movement or action. This process requires cooperative interaction between the central nervous system and the musculoskeletal system, and is thus a problem of information processing, coordination, mechanics, physics, and cognition. Successful motor control is crucial to interacting with the world, not only determining action capabilities, but regulating balance and stability as well.

The organization and production of movement is a complex problem, so the study of motor control has been approached from a wide range of disciplines, including psychology, cognitive science, biomechanics and neuroscience. While the modern study of motor control is an increasingly interdisciplinary field, research questions have historically been defined as either *physiological* or *psychological*, depending on whether the focus is on physical and biological properties, or organizational and structural rules. Areas of study related to motor control are motor coordination, motor learning, and signal processing.

Response to Stimuli

The process of becoming aware of a sensory stimuli and using that information to influence an action occurs in stages, and reaction time of simple tasks can be used to reveal information about these stages. *Reaction time* refers to the period of time between when the stimulus is presented, and the end of the response. *Movement time* is the time it takes to complete the movement. Some of the first reaction time experiments were carried out by Franciscus Donders, who used the difference in response times to a simple reaction task and a choice reaction task to determine the length of time needed process the stimuli and choose the correct response. While this approach is ultimately flawed, it gave rise to the idea that reaction time was made up of a stimulus identification, followed by a response selection, and then the correct movement was carried out. Further research has provided evidence that these stages do exist, but that the response selection period of any reaction time increases as the number of available choices grows, a relationship known as Hick's law.

Closed Loop Control

Most movements that are carried out during day-to-day activity are formed using a continual process of accessing sensory information and using it to more accurately continue the motion. This type of motor control is called feedback control, as it relies on sensory feedback to control movements. Feedback control is a situated form of motor control, relying on sensory information about performance and specific sensory input from the environment in which the movement is carried out. This sensory input, while processed, does not necessarily cause conscious awareness of the action. *Closed loop control* is a feedback based mechanism of motor control, where any act on the environment creates some sort of change that will affect future performance through feedback. Closed loop motor control is best suited to continuously controlled actions, but does not work quickly enough for ballistic actions. Ballistic actions are actions that continue to the end without thinking about it, even when they no longer are appropriate.Because feedback control relies on sensory information, it is as slow as sensory processing. These movements are subject to a speed/accuracy trade-off, because sensory processing is being used to control the movement, the faster the movement is carried out, the less accurate it will become.

Open Loop Control

Some movements, however, occur too quickly to integrate sensory information, and instead must rely on feed forward control. *Open loop control* is a feed forward form of motor control, and is used to control rapid, ballistic movements that end before any sensory information can be processed. In order to best study this type of control, most research focuses on deafferentation studies, often involving cats or monkeys whose sensory nerves have been disconnected from their spinal cords. Monkeys who lost all sensory information from their arms resumed normal behaviour after recovering from the deafferentation procedure. Most skills were relearned, but fine motor control became very difficult.

Coordination

A core motor control issue is coordinating the various components of the motor system to act in unison to produce movement. The motor system is highly complex, composed of many interacting parts at many different organizational levels. Peripheral neurons receive input from the central nervous system and innervate the muscles. In turn, muscles generate forces which actuate joints. Getting the pieces to work together is a challenging problem for the motor system and how this problem is resolved is an active area of study in motor control research.

Reflexes

In some cases the coordination of motor components is hard-wired, consisting of fixed neuromuscular pathways that are called *reflexes.*

Reflexes are typically characterized as automatic and fixed motor responses, and they occur on a much faster time scale than what is possible for reactions that depend on perceptual processing. Reflexes play a fundamental role in stabilizing the motor system, providing almost immediate compensation for small perturbations and maintaining fixed execution patterns. Some reflex loops are routed solely through the spinal cord without receiving input from the brain, and thus do not require attention or conscious control. Others involve lower brain areas and can be influenced by prior instructions or intentions, but they remain independent of perceptual processing and online control.

The simplest reflex is the *monosynaptic reflex* or short-loop reflex, such as the monosynaptic stretch response. In this example, Ia afferent neurons are activated by muscle spindels when they deform due to the stretching of the muscle. In the spinal cord, these afferent neurons synapse directly onto alpha motor neurons that regulate the contraction of the same muscle. Thus, any stretching of a muscle automatically signals a reflexive contraction of that muscle, without any central control. As the name and the description implies, monosynaptic reflexes depend on a single synaptic connection between an afferent sensory neuron and efferent motor neuron. In general the actions of monosynaptic reflexes are fixed and cannot be controlled or influenced by intention or instruction. However, there is some evidence to suggest that the gain or magnitude of these reflexes can be adjusted by context and experience.

Polysynaptic reflexes or long-loop reflexes are reflex arcs which involve more than a single synaptic connection in the spinal cord. These loops may include cortical regions of the brain as well, and are thus slower than their monosynaptic counterparts due to the greater travel time. However, actions controlled by polysynaptic reflex loops are still faster than actions which require perceptual processing. While the actions of short-loop reflexes are fixed, polysynaptic reflexes can often be regulated by instruction or prior experience. A common example of a long loop reflex is the asymmetrical tonic neck reflex observed in infants.

Synergies

A motor *synergy* is a neural organization of a multi-element system that (1) organizes sharing of a task among a set of elemental variables; and (2) ensures co-variation among elemental variables with the purpose to stabilize performance variables. The components of a synergy need not be physically connected, but instead are connected by their response to perceptual information about the particular motor task being executed. Synergies are learned, rather than being hardwired like reflexes, and are organized in a task-dependent manner; a synergy is structured for a particular action and not determined generally for the components themselves. Nikolai Bernstein famously demonstrated synergies at work in the hammering actions of professional blacksmiths. The muscles of the arm controlling the movement of the hammer are informationally linked in such a way that errors and variability in one muscle are automatically compensated for by the actions of the other muscles. These compensatory actions are reflex-like in that they occur faster than perceptual processing would seem to allow, yet they are only present in expert performance, not in novices. In the case of blacksmiths, the synergy in question is organized specifically for hammering actions and is not a general purpose organization of the muscles of the arm. Synergies have two defining characteristics in addition to being task dependent; sharing and flexibility/stability.

"Sharing" requires that the execution of a particular motor task depends on the combined actions of all the components that make up the synergy. Often, there are more components involved than are strictly need for the particular task, but the control of that motor task is distributed across all components nonetheless. A simple demonstration comes from a two-finger force production task, where participants are required to generate a fixed amount of force by pushing down on two force plates with two different fingers. In this task, participants generated a particular force output by combining the contributions of independent fingers. While the force produced by any single finger can vary, this variation is constrained by the action of the other such that the desired force is always generated.

Co-variation also provides "flexibility and stability" to motor tasks. Considering again the force production task, if one finger did not produce enough force, it could be compensated for by the other. The components of a motor synergy are expected to change their action to compensate for the errors and variability in other components that could affect the outcome of the motor task. This provides flexibility because it allows for multiple motor solutions to particular tasks, and it provides motor stability by preventing errors in individual motor components from affecting the task itself.

Synergies simplify the computational difficulty of motor control. Coordinating the numerous degrees of freedom in the body is a challenging problem, both because of the tremendous complexity of the motor system, as well as the different levels at which this organization can occur (neural, muscular, kinematic, spatial, etc.). Because the components of a synergy are functionally coupled for a specific task, execution of motor tasks can be accomplished by activating the relevant synergy with a single neural signal. The need to control all of the relevant components independently is removed because organization emerges automatically as a consequence of the systematic covariation of components. Similar to how reflexes are physically connected and thus do not require control of individual components by the central nervous system, actions can be executed through synergies with minimal executive control because they are functionally connected.

Motor Programs

While synergies represent coordination derived from peripheral interactions of motor components, motor programs are specific, pre-structured motor activation patterns that are generated and executed by a central controller (in the case of a biological organism, the brain). They represent at top-down approach to motor coordination, rather than the bottom-up approach offered by synergies. Motor programs are executed in an open-loop manner, although sensory information is most likely used to sense the current state of the organism and determine the appropriate goals. However, once the program has been executed, it cannot be altered online by additional sensory information.

Evidence for the existence of motor programs comes from studies of rapid movement execution and the difficulty associated with changing those movements once they have been initiated. For example, people who are asked to make fast arm swings have extreme difficulty in halting that movement when provided with a "STOP" signal after the movement has been initiated. Interestingly, this reversal difficulty persists even if the stop signal is presented after the initial "GO" signal but *before* the movement actually begins. This research suggests that once selection and execution of a motor program begins, it must run to completion before another action can be taken. This effect has been found even when the movement that is being executed by a particular motor program is prevented from occurring at all. People who attempt to execute particular movements (such as pushing with the arm), but unknowingly have the action of their body arrested before any movement can actually take place, show the same muscle activation patterns (including stabilizing and support activation that does not actually generate the movement) as when they are allowed to complete their intended action.

Although the evidence for motor programs seems persuasive, there have been several important criticisms of the theory. The first is the problem of storage. If each movement an organism could generate requires its own motor program, it would seem necessary for that organism to possess an unlimited repository of such programs and where these would be kept is not clear. Aside from the enormous memory requirements such a facility would take, no motor program storage area in the brain has yet been identified. The second problem is concerned with novelty in movement. If a specific motor program is required for any particular movement, it is not clear how one would ever produce a novel movement. At best, any new movement would have to be practiced extensively before it could be executed with any success, and at worst no new movements would be possible because no motor program would ever exist. These difficulties have led to a more nuanced notion of motor programs known as generalized motor programs. A generalized motor program is a program for a particular *class* of action, rather than a specific movement. This program is parameterized by the context of the environment and the current state of the organism.

Redundancy

An important issue for coordinating the motor system is the problem of the redundancy of motor degrees of freedom. As detailed in the "Synergies" section, many actions and movements can be executed in multiple ways because functional synergies controlling those actions are able to co-vary without changing the outcome of the action. This is possible because there are more motor components involved in the production of actions than are generally required by the physical constraints on that action. For example, the human arm has seven joints which determine the position of the hand in the world. However, only three spatial dimensions are needed to specify any location the hand could be placed in. This excess of kinematic degrees of freedom means that there are multiple arm configurations that correspond to any particular location of the hand.

Some of the earliest and most influential work on the study of motor redundancy came from the Russian physiologist Nikolai Bernstein. Bernstein's research was primarily concerned with understanding how coordination was developed for skilled actions. He observed that the redundancy of the motor system made it possible to execute actions and movements in a multitude of different ways while achieving equivalent outcomes. This equivalency in motor action means that there is no one-to-one correspondence between the desired movements and the coordination of the motor system needed to execute those movements. Any desired movement or action does not have a particular coordination of neurons, muscles and kinematics which will make it possible, but instead This motor equivalency problem became known as the degrees of freedom problem because it is a product of having redundant degrees of freedom available in the motor system.

Perception in Motor Control

Related, yet distinct from the issue of how the *processing* of sensory information affects the control of movements and actions is the question of how the perception of the world structures action. Perception is extremely important in motor control because it carries the relevant information about objects, environments and bodies which is used in organizing and executing actions and movements. What is perceived and how the subsequent information is used to organize the motor system is a current and ongoing area of research.

Model Based Control Strategies

Most model based strategies of motor control rely on perceptual information, but assume that this information is not always useful, veridical or constant. Optical information is interrupted by eye blinks, motion is obstructed by objects in the environment, distortions can change the appearance of object shape. Model based and representational control strategies are those that rely on accurate internal models of the environment, constructed from a combination of perceptual information and prior knowledge, as the primary source information for planning and executing actions, even in the absence of perceptual information.

Inference and Indirect Perception

Many models of the perceptual system assume indirect perception, or the notion that the world that gets perceived is not identical to the actual environment. Environmental information must go through several stages before being perceived, and the transitions between these stages introduce ambiguity. What actually gets perceived is the mind's best guess about what is occurring in the environment based on previous experience. Support for this idea comes from the Ames room illusion, where a distorted room causes the viewer to see objects known to be a constant size as growing or shrinking as they move around the room. The room itself is seen as being square, or at least consisting of right angles, as all previous rooms the perceiver has encountered have had those properties. Another example of this ambiguity comes from the doctrine of specific nerve energies. The doctrine presents the finding that there are distinct nerve types for different types of sensory input, and these nerves will respond in a characteristic way regardless of the method of stimulation. That is to say, the colour red causes optical nerves to fire in a specific pattern that is processed by the brain as experiencing the colour red. However, if that same nerve is electrically stimulated in an identical pattern, the brain could perceive the colour red when no corresponding stimuli is present.

Forward Models

Forward models are a predictive internal model of motor control that takes the available perceptual information, combined with a particular motor program, and tries to predict the outcome of the planned motor movement.

Forward models structure action by determining how the forces, velocities, and positions of motor components will affect changes in the environment and in the individual. It is proposed that forward models help with the Neural control of limb stiffness when individuals interact with their environment.

Forward models are thought to take motor programs as their input in order to predict the outcome of an action. An error signal is generated when the predictions made by a forward model do not match the actual outcome of the movement, prompting an update of an existing model and providing a mechanism for learning. These models explain why it is impossible to tickle yourself. A sensation is experienced as ticklish when it is unpredictable. However, forward models predict the outcome of your motor movements, meaning the motion is predictable, and therefore not ticklish.

Evidence for forward models comes from studies of motor adaptation. When a person's goal-directed reaching movements are perturbed by a force field, they gradually, but steadily, adapt the movement of their arm to allow them to again reach their goal. However, they do so in such a way that preserves some high level movement characteristics; bell-shaped velocity profiles, straight line translation of the hand, and smooth, continuous movements.

These movement features are recovered, despite the fact that they require startlingly different arm dynamics (i.e. torques and forces). This recovery provides evidence that what is motivating movement is a particular motor plan, and the individual is using a forward model to predict how arm dynamics will change the movement of the arm to achieve particular task level characteristics. Differences between the expected arm movement and the observed arm movement produces an error signal which is used as the basis for learning. Additional evidence for forward models comes from experiments which require subjects to determine the location of an effector following an unvisualized movement

Inverse Models

Inverse models predict the necessary movements of motor components to achieve a desired perceptual outcome. They can also take the outcome of a motion and attempt to determine the sequence of motor commands that resulted in that state. These types of models are particularly useful for open loop control, and allow for specific types of movements, such as fixating on a stationary object while the head is moving. Complimentary to forward models, inverse models attempt to estimate how to achieve a particular perceptual outcome in order to generate the appropriate motor plan. Because inverse models and forward model are so closely associated, studies of internal models are often used as evidence for the roles of both model types in action.

Motor adaptation studies, therefore, also make a case for the existence of inverse models. Motor movements seem to follow predefined "plans" which preserve certain invariant features of the movement. In the reaching task mentioned above, the persistence of bell-shaped velocity profiles and smooth, straight hand trajectories provides evidence for the existence of such plans. The movements which achieve these desired task-level outcomes are estimated by an inverse model. Adaptation therefore proceeds as a process of estimating the necessary movements with an inverse model, simulating with a forward model the outcome of those movement plans, observing the difference between the desired outcome and the actual outcome, and updating the models for a future attempt.

Information Based Control

An alternative to model based control is information based control. Informational control strategies organize movements and actions based on perceptual information about the environment, rather than on cognitive models or representations of the world. The actions of the motor system are organized by information about the environment and information about the current state of the agent. Information based control strategies often treat the environment and the organism as a single system, with action proceeding as a natural consequence of the interactions of this system. A core assumption of information based control strategies is that perceptions of the environment are rich in information and veridical for the purposes of producing actions. This runs counter to the assumptions of indirect perception made by model based control strategies.

Direct Perception

Direct perception in the cognitive sense is related to the philosophical notion of naïve or direct realism in that it is predicated on the assumption that what we perceive is what is actually in the world. James J. Gibson is credited with recasting direct perception as ecological perception. While the problem of indirect perception proposes that physical information about object in our environment is not available due to the ambiguity of sensory information, proponents of direct perception (like Gibson) suggest that the relevant information encoded in sensory signals is not the physical properties of objects, but rather the action opportunities the environment affords.These affordances are directly perceivable without ambiguity, and thus preclude the need for internal models or representations of the world. Affordances exist only as a byproduct of the interactions between an agent and its environment, and thus perception is an "ecological" endeavour, depending on the whole agent/environment system rather than on the agent in isolation.

Because affordances are action possibilities, perception is directly connected to the production of actions and movements. The role of perception is to provide information that specifies how actions should be organized and controlled, and the motor system is "tuned" to respond to specific type of information in particular ways. Through this relationship, control of the motor system and the execution of actions is dictated by the information of the environment. As an example, a doorway "affords" passing through, but a wall does not. How a one might pass through a doorway is specified by the visual information received from the environment, as well as the information perceived about one's own body. Together, this information determines the passability of a doorway, but not a wall. In addition, the act of moving towards and passing through the doorway generates more information and this in turn specifies further action. The conclusion of direct perception is that actions and perceptions are critically linked and one cannot be fully understood without the other.

Behavioural Dynamics

Building on the assumptions of direct perception behavioural dynamics is a behavioural control theory which treats perceptual organisms as dynamic systems that respond to informational variables with actions, in a functional manner. Under this understanding of behaviour, actions unfold as the natural consequence of the interaction between the organisms and the available information about the environment, which specified in body-relevant variables. Much of the research in behavioural dynamics has focused on locomotion, where visually specified information (such as optic flow, time-to-contact, optical expansion, etc.) is used to determine how to navigate the environment Interaction forces between the human and the environment also affect behavioural dynamics as seen in by the Neural control of limb stiffness.

Physiological Basis of Motor Control

Motor Units

Daily tasks, for instance walking to the bathroom, talking one of your friends or eating the dinner all require multiple muscles that innervate body parts to move properly in order to complete specific tasks. Motor units that consist tens, hundreds or even thousands of motor nerves branches are connected to the muscles. In our body, Rectus femoris contains approximately 1 million muscles fibers which are controlled by around 1000 of motor nerves. Within one motor units which can categorized to type I (slow twitch) or Type II fibers (fast twitch), the composition type of the muscle fiber will be consistent (homogeneous); whereas within one muscle, there will be several different combination of two types of motor units (heterogeneous).

There are three primary types of muscle fibers: Type I, Type IIa and Type IIb. As described above, Type I muscle fibers are known as slow twitch oxidative, Type IIa are fast twitch oxidative and Type IIb are fast twitch glycolytic. These three different types of fibers are specialized to have unique functionalities. Type I fibers are described as high endurance but low Force/Power/Speed production, Type IIb as low endurance but high Force/Power/Speed production and Type IIa fibers are characterized in between the two.

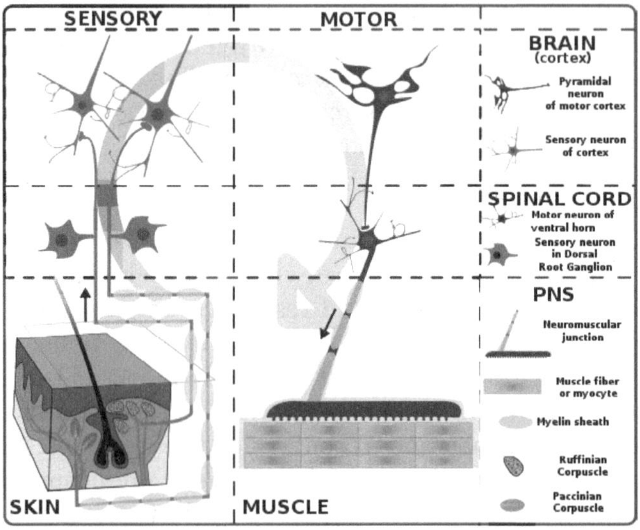

Figure: *Nervous system organization*

Motor units are multiple muscle fibers that are bundled together. When a person wants to move their body, in order to achieve a certain task, the brain instantly sends out an impulse signal that reaches the specific motor unit through the spinal cord. After receiving the signal from the brain, the motor unit contracts muscle fibers within the group thus creating movement. There is no partial firing in the motor unit, meaning, once the signal is detected, all the muscles fibers within the unit contract. However, there are different intensities. Since each motor unit contracts 100% of its fiber once stimulated, types of motor unit that generate different force or speed are significant.

Fiber Type—Contraction Speed—Time to Peak Power—Fatigue

I (slow twitch) ——slow——100 milliseconds————slowly

IIA (fast twitch) ——fast———50 milliseconds———fast

IIB (fast twitch) ——very fast———25 milliseconds—fast

Mechanism and Structure of Motor Unit

Low Threshold Motor Units vs High Threshold Motor Units

For low intensity tasks, smaller motor units with fewer muscle fibers will be used. These smaller motor unites are known as low threshold motor units. They consist of type I fibers that contract much slower and thus provide less force for daily basic movement such as typing on the keyboard. For more intense tasks, motor units containing Type II muscle fibers will be utilized. These fast twitch motor units are known as high threshold motor units. The major difference between low threshold motor units (slow twitch motor unit) and high threshold motor units (fast twitch motor unit) is that high threshold motor units control more muscle fibers and contain larger muscle fibers, in comparison to low threshold motor unit. On the other hand, the main difference between the slow twitch muscle fiber (Type I) and fast twitch muscle fiber (Type II) has the same theory of the size deviations.

The order of recruitment of motor unit

During an activity of lifting heavy objects such as working out with a dumbbell, not only does low threshold motor units, but also the high threshold motor units are recruited to compensate forces required in addition to just holding a fork, in which the energy created by the low threshold motor units is sufficient to complete the job. When giving a job, the body first recruits the slow-twitch motor units following by recruit more a more fast-twitch motor units as forces required to complete the movements increase. Thus, when the body has to carry an extremely massive object, it would recruited all the available motor units to contract for the particular muscle that has been used.

```
                 High|
                     |                                   _____________
Force required       |                                  /
                     |                                 |
                     |                                 |
                     |                       __________|_____________
                     |            ________|______________________________
                  Low|_________|________________________________________
                               ?        ?              ?                 Time
               Type I Recruit first  Type II A     Type IIB
```

Fiber Versus Nerves

The type of fiber (type I vs Type II) is controlled by the nervous system. The brain is the central information centre that sent out the signal to the nerves, that the nerves control and connect the motor units. For two different motor units present, the body adopt it with two different nerves to control them. Fast twitch motor units are controlled by fast twitch nerves while Slow twitch motor units are controlled by slow twitch nerves.

In the laboratory, a nerve from a motor unit that is connected to a slow twitch muscle fiber was replace with a nerve that are designate for a fast twitch fiber. Amazingly, the slow twitch fiber behaved identically as a fast twitch fiber. In contrast, if the process was reverse, the fast twitch fiber performed as a slow twitch fiber as well. However, the nerves can not possibly transform from fast motor nerves into slow motor nerves and vice versa.

In many sports movements, the duration of certain action usually are within 200 milliseconds, and from the above charts, time to peak power of the individual muscle fibers of each type (I,IIA,IIB) have sufficient time to reach peak power production. This brings out a question of what is the superiority of having more Type II fibers?

This can be documented when you analyze a large group of athletes for vertical jump performance and their execution for a vertical jump. Athletes with more fast-twitch fibers (Type II) change direction quicker during their movement such as left to right direction and they tend to use less knee bend. These results can be confirmed by muscle biopsy and even by special force-plate analysis. This does not mean that athletes with lower fast-twitch fiber cannot jump higher, but they tend to do it a little slower and with a deeper knee bend.

Although having a high percentage of Type II fibers give a person more quicker movement, there is little doubt that the nervous system and the brain is more important on affecting the performance.

Muscle Motor Unit Recruitment

The majority of the time, the real limit to athletes performance is the number of motor units their nervous system can recruit in the short period of time and the amount of forces (size of the muscle fibers) provided of those motor units. The performance is rarely affect by the type of muscle fiber (slow twitch or fast) that constructed to motor units. The nervous system determines the degree of motor unit activated in sport like activities.

It normally takes 0.4 - 0.6 seconds for the nervous system to activate available muscle motor units to contract, same length of time demonstrating maximum strength or force. However, vertical jump activity only takes 0.2 seconds to perform. Therefore the factor of determining the performance is within 0.2 seconds, how many available muscle motor units can be recruited for contraction, yet how much fast twitch fibers in the body. as the result, an athlete lacking fast twitch fibers has better control of nervous system that recruited all the fast-twitch fibers in the body, the athlete tend to have superior performance in comparison to the athlete with a less control of nervous system while having greater number of fast twitch fibers.

From above, people can considerably increase their strength without increase the size of their muscle, because the body becomes more efficient at muscle recruitment and firing synchronization.

Biological Homeostasis

All living organisms depend on maintaining a complex set of interacting metabolic chemical reactions. From the simplest unicellular organisms to the most complex plants and animals, internal processes operate to keep the conditions within tight limits to allow these reactions to proceed. Homeostatic processes act at the level of the cell, the tissue, and the organ, as well as for the organism as a whole.

Principal Homeostatic processes include the following:

- "Warm-blooded" (endothermic) animals (mammals and birds) maintain a constant body temperature, whereas ectothermic animals (almost all other animals) exhibit wide body temperature variation. An advantage of temperature regulation is that it allows an organism to function effectively in a broad range of environmental conditions. For example, ectotherms tend to become sluggish at low temperatures, whereas a co-located endotherm may be fully active. That thermal stability comes at a price, since an automatic regulation system requires additional energy. If the temperature rises, the body loses heat by sweating or panting, via the latent heat of evaporation. If it falls, this is counteracted by increased metabolic action, by shivering, and—in fur- or feather-coated creatures—by thickening the coat.

- Regulation of the pH of the blood at 7.365 (a measure of alkalinity and acidity).
- All animals also regulate their blood glucose concentration. Mammals regulate their blood glucose with insulin and glucagon. The human body maintains glucose levels constant most of the day, even after a 24-hour fast. Even during long periods of fasting, glucose levels are reduced only very slightly. Insulin, secreted by the beta cells of the pancreas, effectively transports glucose to the body's cells by instructing those cells to keep more of the glucose for their own use. If the glucose inside the cells is high, the cells will convert it to the insoluble glycogen to prevent the soluble glucose from interfering with cellular metabolism. Ultimately this lowers blood glucose levels, and insulin helps to prevent hyperglycemia. When insulin is deficient or cells become resistant to it, diabetes occurs. Glucagon, secreted by the alpha cells of the pancreas, encourages cells to break down stored glycogen or convert non-carbohydrate carbon sources to glucose via gluconeogenesis, thus preventing hypoglycemia.
- The kidneys are used to remove excess water and ions from the blood. These are then expelled as urine. The kidneys perform a vital role in homeostatic regulation in mammals, removing excess water, salt, and urea from the blood.
- If the water content of the blood and lymph fluid falls, it is restored in the first instance by extracting water from the cells. The throat and mouth become dry, so that the symptoms of thirst motivate the animal to drink.
- If the oxygen content of the blood falls, or the carbon-dioxide concentration increases, blood flow is increased by more vigorous heart action and the speed and depth of breathing increases.
- Sleep timing depends upon a balance between homeostatic sleep propensity, the need for sleep as a function of the amount of

time elapsed since the last adequate sleep episode, and circadian rhythms that determine the ideal timing of a correctly structured and restorative sleep episode.

- Personality traits are often conceptualized as a person specific setpoint level around which mood states fluctuate in time.

Control Mechanisms

All homeostatic control mechanisms have at least three interdependent components for the variable being regulated: The receptor is the sensing component that monitors and responds to changes in the environment. When the receptor senses a stimulus, it sends information to a "control centre", the component that sets the range at which a variable is maintained. The control centre determines an appropriate response to the stimulus. The control centre then sends signals to an effector, which can be muscles, organs, or other structures that receive signals from the control centre. After receiving the signal, a change occurs to correct the deviation by depressing it with negative feedback.

Negative Feedback

Negative feedback mechanisms consist of reducing the output or activity of any organ or system back to its normal range of functioning. A good example of this is regulating blood pressure. Blood vessels can sense resistance of blood flow against the walls when blood pressure increases. The blood vessels act as the receptors and they relay this message to the brain. The brain then sends a message to the heart and blood vessels, both of which are the effectors. The heart rate would decrease as the blood vessels increase in diameter (known as vasodilation). This change would cause the blood pressure to fall back to its normal range. The opposite would happen when blood pressure decreases, and would cause vasoconstriction.

Another important example is seen when the body is deprived of food. The body would then reset the metabolic set point to a lower than normal value. This would allow the body to continue to function, at a slower rate, even though the body is starving. Therefore, people depriving themselves of food while trying to lose weight would find it easy to shed weight initially and much harder to lose more after. This is due to the body's readjusting itself to a lower metabolic set-point to allow the body to survive with its low supply of energy. Exercise can change this effect by increasing the metabolic demand.

Another good example of negative feedback mechanism is temperature control. The hypothalamus, which monitors the body temperature, is capable of determining even the slightest variation of normal body temperature (36.5 degrees Celsius). Response to such variation could be stimulation of glands that produce sweat to reduce the temperature or signaling various muscles to shiver to increase body temperature.

Both feedbacks are equally important for the healthy functioning of one's body. Complications can arise if any of the two feedbacks are affected or altered in any way.

Homeostatic Imbalance

Many diseases involve a disturbance of homeostasis. As the organism ages, the efficiency in its control systems becomes reduced. The inefficiencies gradually result in an unstable internal environment that increases the risk of illness, and leads to the physical changes associated with aging.

Certain homeostatic imbalances, such as high core temperature, a high concentration of salt in the blood, or low concentration of oxygen, can generate homeostatic emotions (such as warmth, thirst, or breathlessness), which motivate behaviour aimed at restoring homeostasis (such as removing a sweater, drinking or slowing down).

Bibliography

Anderson, JA; Rosenfeld, E.: *Neurocomputing: Foundations of Research*, The MIT Press, Cambridge, Mass., 1988.

Atkins, Peter W. and Julio de Paula: *Physical Chemistry*, Wiley-VCH, Weinheim, 2006.

Berg, Jeremy M.; Tymoczko, John L.; Stryer, Lubert: *Biochemistry*, W. H. Freeman, New York, 2007.

Brock, William H.: *Viewegs Geschichte der Chemie*, Vieweg, Braunschweig, 1997

Brückner, Reinhard, *Reaktionsmechanismen,* Spektrum Akademischer Verlag, München, 2004.

Bugg, Tim: *Introduction to Enzyme and Coenzyme Chemistry*, Blackwell Publishers, Cambridge, MA, 2004.

Bullock, TH; Orkand, R; Grinnell, A.: *Introduction to Nervous Systems*, W. H. Freeman, San Francisco, 1977.

Campbell, Neil A.; Brad Williamson; Robin J. Heyden: *Biology: Exploring Life*, Pearson Prentice Hall, Boston, Massachusetts, 2006.

Cornish-Bowden, Athel: *Fundamentals of enzyme kinetics,* Portland Press, London, 2004.

Danson, Michael; Eisenthal, Robert: *Enzyme assays: a practical approach,* Oxford University Press, London, 2002.

Elschenbroich, Christoph: *Organometallchemie*, Vieweg Teubner Verlag,Wiesbaden, 2008.

Fersht, Alan: *Structure and mechanism in protein science*, W.H. Freeman, San Francisco, 1999.

Ganong, WF: Review of Medical Physiology, Appleton and Lange, Norwalk, Conn., 1991.

JM Berg, JL Tymoczko, L Stryer: *Biochemistry*, WH Freeman, New York, 2003.

Kandori, Hideki: *Retinal Binding Proteins*, Wiley-VCH, New York, 2006.

Karlson, Peter; Doenecke, Detlef; Fuchs, Georg; Gerok, Wolfgang: *Karlson Biochemistry and Pathobiochemistry,* Thieme, Stuttgart, Germany, 2005.

Keynes, RD; Aidley, DJ.: *Nerve and Muscle,* Cambridge University Press, Cambridge, 1991.

Koolman, J.; K. H. Roehm: *Color Atlas of Biochemistry,* Georg Thieme Verlag, Stuttgart, Germany, 2005.

Latscha, Hans Peter; Kazmaier, Uli; Alfons, Klein Helmut: *Organische Chemie: Chemie-basiswissen II,* Springer, New York, 2008.

Lechner, Manfred; Gehrke, Klaus; Nordmeier, Eckhard: *Macromolecular Chemistry,* Birkhäuser, Basel, 2003.

Liang-Hong Guo; H. Allen; O. Hill: *Direct Electrochemistry of Proteins and Enzymes,* Academic Press, San Diego, 1991.

Lodish H Berk A, Matsudaira P, Kaiser CA, Krieger M, Scott MP, Zipursky SL, Darnell J.: *Molecular Cell Biology,* WH Freeman, New York, 2004.

March, Jerry: *Advanced Organic Chemistry: Reactions, Mechanisms, and Structure,* Wiley, New York, 1985.

Marye Anne Fox; James K. Whitesell (2004). Organic chemistry (Third ed.). Jones & Bartlett. p. 699. ISBN

S. R. Hartshorn: *Aliphatic Nucleophilic Substitution,* Cambridge University Press, London, 1973.

Schmidt-Nielsen, K.: *Animal Physiology: Adaptation and Environment,* Cambridge University Press, Cambridge, 1997.

Schwann, HP.: *Biological Engineering,* McGraw-Hill, New York, 1969.

Spanswick, RM; Lucas, WJ; Dainty, J.: *Plant Membrane Transport: Current Conceptual Issues,* Elsevier Biomedical Press, Amsterdam, 1980.

Stevens, Lewis; Price, Nicholas C.: *Fundamentals of enzymology: the cell and molecular biology of catalytic proteins,* Oxford University Press, Oxford, 1999.

Waxman, SG,: *Molecular Neurology,* Elsevier Academic Press, Burlington, Mass., 2007.

Wiberg, Egon, Wiberg, Nils and Holleman, Arnold Frederick: *Inorganic chemistry,* Academic Press, Waltham, Massachusetts, 2001.

Worden, FG; Swazey, JP; Adelman, G,: *The Neurosciences, Paths of Discovery,* The MIT Press, Cambridge, Mass., 1975.

Wrighton, Mark S.; Ebbing, Darrell D.: *General chemistry,* Houghton Mifflin, Boston, 1993.

Index

H

I

L

M

N

O

P

R

S

T

V

X

❑❑❑